PASTORAL PROPERTIES
OF AUSTRALIA

PASTORAL PROPERTIES
OF AUSTRALIA

PETER TAYLOR

GEORGE ALLEN & UNWIN
Sydney   London   Boston

First published in 1984
George Allen & Unwin Australia Pty Ltd
8 Napier Street, North Sydney NSW 2060 Australia

George Allen & Unwin (Publishers) Ltd
18 Park Lane, Hemel Hempstead Herts HP2 4TE England

Allen & Unwin Inc.
9 Winchester Terrace, Winchester Mass 01890 USA

National Library of Australia
Cataloguing-in-Publication entry:

Taylor, Peter, 1936–.
Pastoral properties of Australia.
Bibliography.
Includes index.
ISBN 0 86861 768 7.
1. Sheep ranches — Australia.   2. Ranches —
Australia.   I. Title.
635.01′0994

Set in Goudy Old Style Condensed by
Graphicraft Typesetters, Hong Kong
Printed in Singapore
Designed by Judy Hungerford

Title page: Dawn at Tully River Station, Northern Queensland

CONTENTS

ACKNOWLEDGEMENTS

Many people helped me to write this book and I am grateful for their time and expertise, both of which they gave unstintingly.

I would particularly thank Doug Mactaggart, a great cattleman and a great Queenslander, in Brisbane; and Ralph Storey, for many years Secretary of the Australian Association of Stud Merino Breeders and now in retirement at Bowral. They gave me an enormous amount of help when I most needed it — at the beginning — and continued to help me throughout the long preparation of this book.

I would also like to thank Arthur Bassingthwaighte, AO. and Len Davidson of King Ranch; John Conrick; Noreen Coulton, Dr Peter Booth and Ken Bennett of the Australian Wool Corporation; Ernie Ecob, NSW State Secretary, the Australian Workers Union; Tim Emanuel; Suzanne Falkiner; Peter Freeman of the Australian Heritage Commission; G. Higginson of Elders Pastoral; Fred Hockey, the School of the Air, Alice Springs; Jim and Doug Lowden; Tara McCarthy and Colin Munro of the A.B.C.; Ken Nicholas and John Hepworth of the Royal Flying Doctor Service; Bill Norton and John Amstrong of Stanbroke Pastoral Company; Margaret Olds; Trevor Schmidt of the Australian Agricultural Company; Helen Tolcher; Sue Wagner; Roger Ward of George Allen & Unwin; James S. White; Owen Williams of CSIRO, Canberra.

I must also thank the Mitchell Library, Sydney; the La Trobe Library, Melbourne; South Australian Archives, Adelaide; Australian Archives, Darwin; the National Library of Australia, Canberra; the State Library of Tasmania, Hobart; the Oxley Memorial Library, Brisbane; and the J. S. Battye Library, Perth. I also acknowledge my debt to those who have written books on the history of individual stations — although not numerous, they made my task much easier than it would have been.

I am, of course, particularly grateful to the owners of the properties described in this book for allowing me to visit them. Their help, encouragement and hospitality were simply invaluable and this book could not have been written without them.

Finally, but by no means least, I would thank the men, women and children who live and work on the properties I visited. Some are mentioned by name, and I hope those who are not will forgive me, for the list would run to hundreds. I thank them all for their help, courtesy and friendship which never failed, no matter how stupid I was or how busy they were. I would do more than thank them, for that is not enough. Rather, I offer this book to them as a token of gratitude and admiration, and of my humility, for they taught me much.

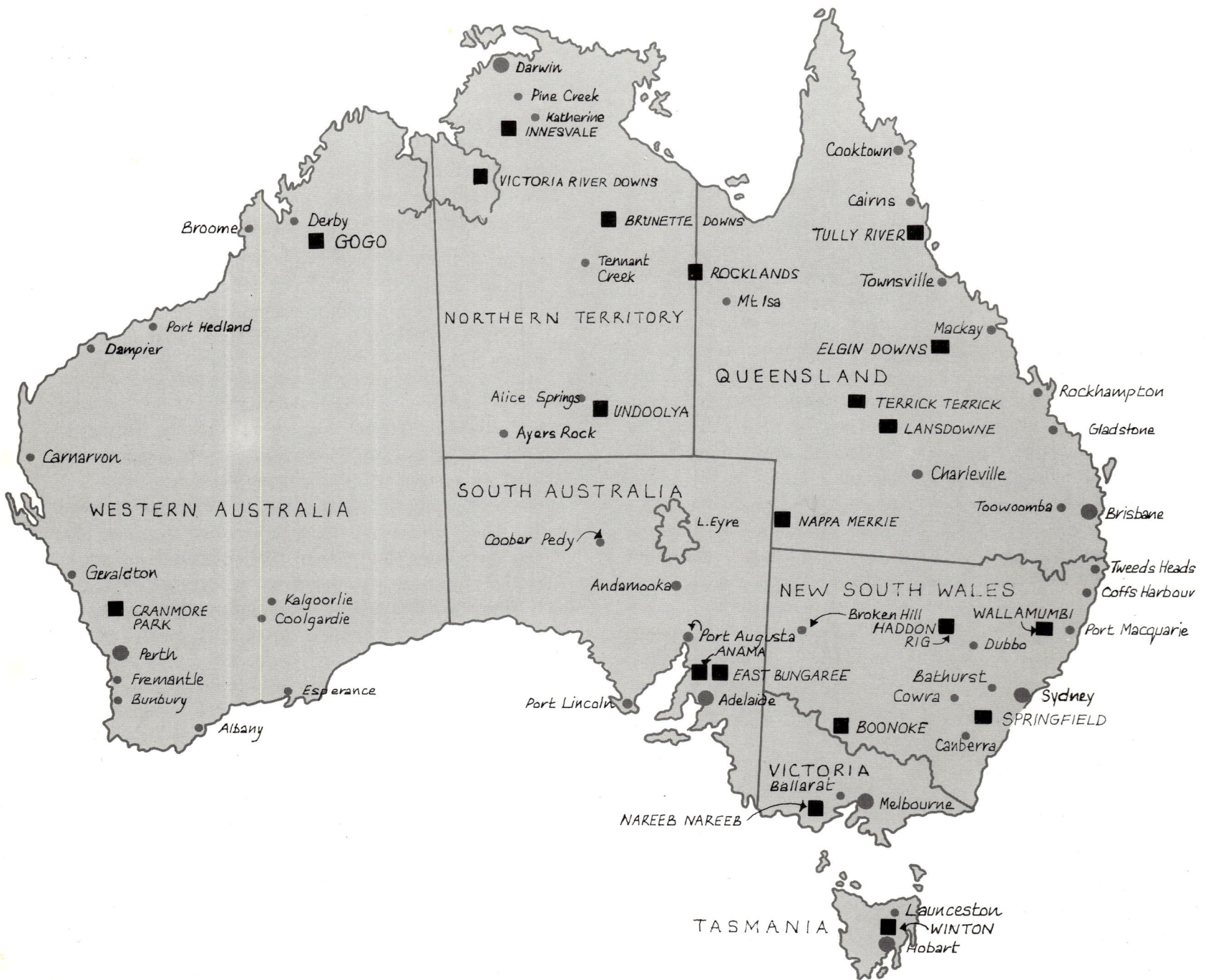

Darwin
Pine Creek
Katherine
INNESVALE

Cooktown

Cairns

TULLY RIVER

VICTORIA RIVER DOWNS

BRUNETTE DOWNS

Townsville

Broome
Derby
GOGO

Tennant
Creek

ROCKLANDS
Mt Isa

Mackay

Port Hedland
Dampier

ELGIN DOWNS

NORTHERN TERRITORY

QUEENSLAND

Rockhampton

Carnarvon

Alice Springs
UNDOOLYA

TERRICK TERRICK
LANSDOWNE

Gladstone

Ayers Rock

WESTERN AUSTRALIA

SOUTH AUSTRALIA

Charleville

Toowoomba

Brisbane

L. Eyre

Coober Pedy

NAPPA MERRIE

Tweeds Heads

Geraldton

CRANMORE
PARK

Andamooka

NEW SOUTH WALES

Coffs Harbour

Kalgoorlie
Coolgardie

Broken Hill
HADDON
RIG

WALLAMUMBI

Port Macquarie

Perth
Fremantle
Bunbury

Port Augusta
ANAMA

Dubbo

Esperance

EAST BUNGAREE

Bathurst
Cowra

Port Lincoln

Adelaide

Sydney
SPRINGFIELD

Albany

BOONOKE

Canberra

VICTORIA
Ballarat

NAREEB NAREEB

Melbourne

TASMANIA

Launceston
WINTON
Hobart

INTRODUCTION

This book describes twenty pastoral properties in Australia — I have tried to show how they came about, what they are like now, what they do, and how the people on them work and live.

Although Australia is justifiably famous for its sheep and cattle stations, some of which are huge by any standard, few Australians, and even fewer overseas visitors, ever have the chance to see them. Many are remote from the main centres of population and in any case none of them sets out to attract tourists. They are working stations and the people on them always have much to do.

This book, then, is for those who are curious about such places but who have little or no chance of ever seeing them.

Selecting the properties to be included was very difficult. It would have been easy to pick the oldest, the biggest and the most famous, but such a selection would have shown a very limited aspect of the huge variety of properties that make up the Australian pastoral industry. Instead, they have been selected to give a much broader view of that variety — to show how different they are, rather than to show the similarities. But they do have some things in common: the sheep properties are all merino studs and the cattle properties are all running beef cattle.

The stations were selected either because they are typical of a geographical area, or because their contribution is in some way typical of different techniques. Some are big and famous, some are historical, and some are neither, but together they make a more comprehensive picture than one based on romance alone.

I visited each of these properties between November 1982 and August 1983 and travelled some 30,000 kilometres in the process. During that time many properties in the south and east of Australia were very badly affected by one of the worst droughts on record. They did not look their best, and it is a tribute to their owners that they allowed me to photograph them under such circumstances. Their view, and mine, was that we should show how these places are in the bad times, what they are like when they are struggling to keep stock alive and when resources are being strained to the limit. If in some cases the country looks barren or the stock not up to standard, that is why. Country people will understand.

I spent about a week on each property. During that time I saw as much of the country as I could, talked to as many people as I could, and spent the working day with them whenever I could. None of the properties made any special arrangements for me, nor did I ask for any. Nothing was 'set up' to show things in a favourable light. I simply photographed and wrote about what was going on around me. If I had not been there, they would have been going on just the same. If some of the horsemen do not have the style of the show ring, for example, it is because they were not in a show ring. They were working, usually very hard, and style was not uppermost in their minds.

I thought long and hard about the use of metrics in this book and in the end I decided to use the measurements country people use. I never once heard anybody describe land in hectares or rainfall in millimetres, so I have adopted their use of acres for land and inches for rainfall. This might infuriate those who would have us convert completely to metrics but I see little merit in writing in a way not readily understood by those likely to read it. For those who *are* totally converted, an acre is 0.4 ha and an inch is 25.4 mm. Other measurements are in metrics because they were in common use.

This book has taken eighteen months to write. Much of that time was spent travelling to parts of Australia that I probably would not otherwise have seen, and talking to people I would have been unlikely to meet anywhere else. It was a memorable experience. The country impressed me as one of the most beautiful in the world, and the people impressed me even more.

I do not know if this is the 'real' Australia and these the 'real' Australians, as some suggest. I am not even sure what they mean. But had I discovered *that* Australia and *those* Australians when I was younger, I think I might not have come back.

PETER TAYLOR
Sydney 1984

AUSTRALIAN WOOL

Although Australia runs sheep for meat, and in considerable quantities, the sheep that produce wool are much more famous. When school children are told that Australia rides on the sheep's back it is no more than the truth, and if the back is supported by four edible legs it is not of much importance.

So, if sheep means wool in Australia, wool means merino. Certainly there are other breeds. There are the famous sheep of Britain and there are breeds that have been produced by crossing them with merinos. Most of these have more to do with meat than wool. Although they are important, it does not alter the fact that by far the greatest number of sheep in Australia are merinos.

Why, then, is the merino so important? The answer is not only the quality of the wool it produces, but also the amount. On a sheep the wool grows through follicles in the skin. Each primary follicle is surrounded by a number of secondary follicles through which the wool grows. Most sheep have between three and seven secondary follicles around each primary. But the merino has between twenty and twenty-five secondary follicles. Not only is the wool finer, but there are about five times as many fibres as on other breeds. That is why it is the supreme wool producing animal. And by a lucky accident of history, not only was this animal introduced to Australia, it also proved to be ideally suited to its conditions.

Not much is known about the origin of the merino as a breed until it appeared in Spain in the twelfth century. They were introduced there by north African Berbers called Beni-Merines, hence its name. Its ability to produce fine wool was soon recognised. As flocks grew they were jealously guarded by the Spaniards: merinos were kept under royal patronage and very few left the country.

Eventually, however, and fortunately for Australia, some did leave. In 1765 the King of Spain gave a hundred merino rams and two hundred ewes to his cousin the Elector of Saxony, part of what is now Germany. This was followed by a similar gift in 1778 and by 1802 there were about four million merinos in Saxony, all of the finest breed.

King Louis of France was also able to obtain a number of merinos from Spain and established them on an experimental farm at Rambouillet in 1786.

As the leading manufacturer of textiles, it was inevitable that Britain also became anxious to import some of these remarkable animals, if only to reduce the cost of importing wool from those lucky enough to have them. But relations between Britain and Spain at that time were far from friendly and the regulations in Spain that prohibited exports were strictly enforced. It was too delicate a matter for King George III, who was keen to acquire some, so in 1787 Sir Joseph Banks acquired two rams and four ewes from Portugal by what was described as 'gradual means'.

In spite of Sir Joseph's enthusiasm for the new colony of New South Wales, the first Australian merinos did not come from him. In 1797 two ships from Sydney, the *Reliance* and the *Supply*, put into Cape Town to buy cattle and other

An eighteenth century pure Spanish merino ram with a fleece of about 3 kg.
REPRODUCED BY PERMISSION OF THE NATIONAL LIBRARY OF AUSTRALIA.

supplies for the colony. There they found that Mrs Gordon, widow of a British Governor, was disposing of a flock of merinos descended from sheep which were a gift from the King of Spain.

She gave three to Philip King and three to Colonel Patterson, who were returning to England from New South Wales, and she offered to sell the rest to the New South Wales Commissary. He refused them on the grounds that he had no authority to buy them, so instead they were bought by the captains of the ships, Captain Kent and Captain Waterhouse, who each bought thirteen animals for four guineas a head.

Kent's sheep were loaded on to the *Supply*, but most died during the voyage to New South Wales. The few survivors were given to the Governor in Sydney but these died too. Fortunately those brought in the *Reliance* by Waterhouse survived and they were sold to important landowners in New South Wales, among them the Reverend Samuel Marsden and John Macarthur, both of whom already had sizeable flocks of Cape sheep that had been imported earlier from the Cape.

Macarthur had also previously bought sixty Bengal ewes and lambs and had since added two Irish ewes and a ram. When he crossed these he found that the lambs produced a fleece of hair and wool that was still coarse, but which was much better than previous fleeces. His acquisition of four merino ewes and two rams from Waterhouse now gave him the chance to produce much finer wool.

Whilst his claim to be the father of the Australian wool industry is not so strong as was once thought (and his wife Elizabeth is now known to have played a considerable part), there is no doubt that his search for a product that could be sent overseas led him to the firm conclusion that this product should be wool. It was a cause he supported for the rest of his life.

In 1801 Macarthur was sent back to England to face a court martial after fighting a duel with his commanding

A ram imported from Saxony in 1824. REPRODUCED BY PERMISSION OF THE NATIONAL LIBRARY OF AUSTRALIA.

This photograph, which was taken in the early 1860s, shows the ram 'Napoleon', which was one of three rams used by George Peppin to start his famous flock at Wanganella. The boy is George F. Peppin.

officer. Most men would have seen it as the end of a promising career, but instead Macarthur saw it as an opportunity. It gave him a chance to talk to people in London about the future of Australian wool as he saw it. He told them he had about 4,000 sheep and that in twenty years they would have increased to the point where they would be able to supply most of England's need for fine wool.

As England was at that time importing nearly £2 million worth of wool each year (merinos not having been successful there), it is not surprising that his comments were listened to with interest. The samples of wool that he showed also seemed to confirm that Australian wool might indeed soon be as good as any from Europe.

His visit to England coincided, by happy chance, with the first public sale of some of the King's famous merinos. This took place at Kew in August 1804 and was attended by about fifty breeders. Macarthur bought the first lot, a ram, for £6 15s, but the bidding may not have been very keen as it was 'labouring under a temporary privation of sight'.

By the end of the sale Macarthur had bought seven rams and one ewe, in spite of an old Act that decreed that the sheep could not be exported. But for a country that had itself obtained sheep by 'gradual means' this was not a major problem. A dispensation was granted and in June 1805 Macarthur and his sheep sailed into Sydney Harbour on board the *Argo*, which fittingly carried a golden fleece as its figurehead. Five rams and the ewe survived the voyage and their descendants were known for many years as the 'Argo lot'.

Macarthur returned to Sydney with more than a few sheep, for he had also been given permission in London to select a block of land and to devote himself to the production of fine wool. He chose land in an area south-west of Sydney and called his property Camden Park.

At that time the major supplier of fine wool to the English textile mills was Saxony, and she represented a formidable competitor for the new Australian industry. Saxony had put to good use the present from the King of Spain and since 1765 had successfully developed big flocks of merinos. Their wool set the standard in England and even an increase in import duty there did little to reduce the flow, let alone break their monopoly.

They probably took little notice when a London buyer paid 10/4 a pound for Macarthur's first bale of fine wool, which he had shipped in 1807. At this time most Australian wool was still coarse and dirty and few took it seriously in London. In Saxony it was simply irrelevant.

Although the emphasis in Australia at that time was on meat production, Macarthur and a few other breeders continued the slow development of the fine-wooled merino and used it to improve the quality of other sheep. By 1815 they had bred out the hairiness that had been a feature of the non-merino breeds and by 1820 they were exporting about half their annual clip.

When Macarthur won a gold medal in England for producing wool as fine as that from Saxony, and the British government reduced the tariff on wool from Australia, it was the start of a trade war that was to be of vital importance to the new colony.

In spite of their domination of the market, Saxony had problems. Because they *had* dominated, they were not prepared for vigorous competition. Also, in producing wool of increasingly fine quality they had produced an animal that lacked natural vigour and which produced only a small amount of wool. They had, in fact, bred an animal that was not equipped to compete in the mass apparel market which was developing in the middle of the Industrial Revolution.

Australia had problems too. The country was a long way from the British market and for a long time British buyers remained sceptical about the quality of Australian wool. But Australian growers had vigorous animals and a new industry, and before long exports grew. By 1836 they were exporting more than 1.5 million kilograms of wool and by the 1840s the trade war had been won.

A merino fleece — cornerstone of the Australian wool industry.

Faced with falling prices as a result of the competition, sheep numbers in Saxony fell. In 1845 German manufacturers started to import wool from Australia. With the trade war over, the way was now open for Australian growers to concentrate on wool to supply the growing needs of the industrial cities of Europe.

So far, most fine wool had come from the descendants of Macarthur's Camden flock, which had been crossed with other merinos from Saxony and France on a more or less haphazard basis. Now, as the demand for fine wool decreased in favour of stronger wool suitable for a wide range of garments, further developments were needed. In 1858 the Peppin brothers started to develop a flock which was to result in a new type of animal which could produce exactly the type of wool now required — the medium wool Peppin merino.

They established their stud, called Wanganella, near Deniliquin in 1861 and started to breed from merinos from France and Spain. Records are scanty, but one ram in particular — a Rambouillet ram called Emperor — had a profound influence. It is now acknowledged as being perhaps the most important ram in the history of Australian wool.

Eventually the Peppins produced an outstanding merino that could thrive in dry areas, that could look after itself much of the time, and which produced a heavy fleece of

Inside the ram shed at East Bungaree, South Australia.

14

A team of blade-shearers at Canowie, South Australia, about 1870. The click of the shears occurred when the shearer used the full length of the blades and brought the 'knockers' together. Some studs still use blades to shear their top rams.

medium-strength wool. There was a great deal of natural grease in the wool and this protected it from the worst of the climate and gave it a distinctive creamy colour.

Today nearly three-quarters of the Australian sheep population are direct descendants of Peppins' merinos. A stud ram will often produce 20 kg of greasy wool, compared to the 2 kg produced by early Australian sheep. The Peppin merino is common in Queensland, New South Wales and northern Victoria, but it is adaptable enough to survive also in the heavier rainfall areas of southern Victoria and Tasmania. So important is it that even now flocks are often described simply as Peppin or non-Peppin.

Whilst the Peppins were developing this famous sheep, other people were experimenting too. In 1866 a few rams were imported from America to provide the first contact with the original Spanish blood for nearly sixty years. The dark-tipped merino from Vermont was thought to be very impressive as it carried an enormous fleece on its wrinkled skin. The trickle became a flood and by the 1890s they were being imported in vast numbers.

But breeders soon realised that the wrinkles had serious disadvantages when it rained in a dry country. By 1905, after a series of massive fly-strikes, the love affair was over. By then many studs had been ruined by this blood as breeders rushed for the benefits without waiting to see if the animal was suitable for the climate. There was nothing wrong with the Vermont strain except that it was in the wrong place and breeders learnt, if they needed to, that fashion could be dangerous. There is now no trace of the Vermont on today's Australian merino, although wrinkles are still present in some strains.

Meanwhile, other strains of merinos were being developed for those parts of the country where the Peppin was less suitable, and this produced three quite distinct strains of merino.

The South Australian merinos were bred to succeed in the very arid areas of that State, where rainfall is less than 10 inches a year and saltbush makes up most of the vegetation. This merino is the biggest of the strains and has a body that is longer, taller and heavier than the Peppin. The skin is less wrinkled and the wool is stronger. The South Australian merino is found in parts of Western Australia, Queensland and New South Wales as well as in its home State and it has been extremely successful in those areas.

The second strain is the Spanish merino, which is a direct descendant from Spanish blood. They are about the same size as the Peppin and produce about the same amount of wool. They are successful in the drier areas but they are the least numerous of the merino strains.

Finally there is the Saxon merino which is still found in the areas of higher rainfall in the southern States, where the Peppin is less successful. Indeed, it is very different to the Peppin. It is much smaller and produces a fleece of only about 5 kg. But its wool is of the very finest quality. It is soft and white and is used for the highest grade of cloths. In 1973 a bale of Tasmanian wool from this strain set a new world record when it sold for $42 per kg.

So, after nearly two hundred years of development, the Australian wool industry stands supreme. The growing use of technology has produced a national flock of nearly 140 million sheep, including those breeds such as Corriedale and Polwarth that have been produced by crossing British breeds and which are used to produce meat as well as wool.

But it is the Australian merino that is truly outstanding. The merino of today carries a fleece that would have crippled the poor sheep that Macarthur knew. It has been produced by generations of skilled Australian sheep breeders who had to develop their own techniques in the heat and isolation of the Australian bush. And in the process they established a tradition for perfection that is still very much alive today.

WINTON

On the plains of the Macquarie River 15 kilometres from Campbell Town in central Tasmania is Winton, a quite remarkable sheep stud. It is remarkable for two reasons. One is that it is the oldest surviving stud in Tasmania (and probably Australia), and the other is that the sheep are some of the purest Saxon merinos to be found anywhere in the world.

Consisting of 8,250 acres, the property has a narrow frontage on the Macquarie River. From there it runs northeast to the main road between Launceston and Campbell Town, whilst another road, from Campbell Town to Longford, runs through the property a kilometre or so from the river. The country is gentle and undulating, becoming flatter near the river. Much of it is open country although a part

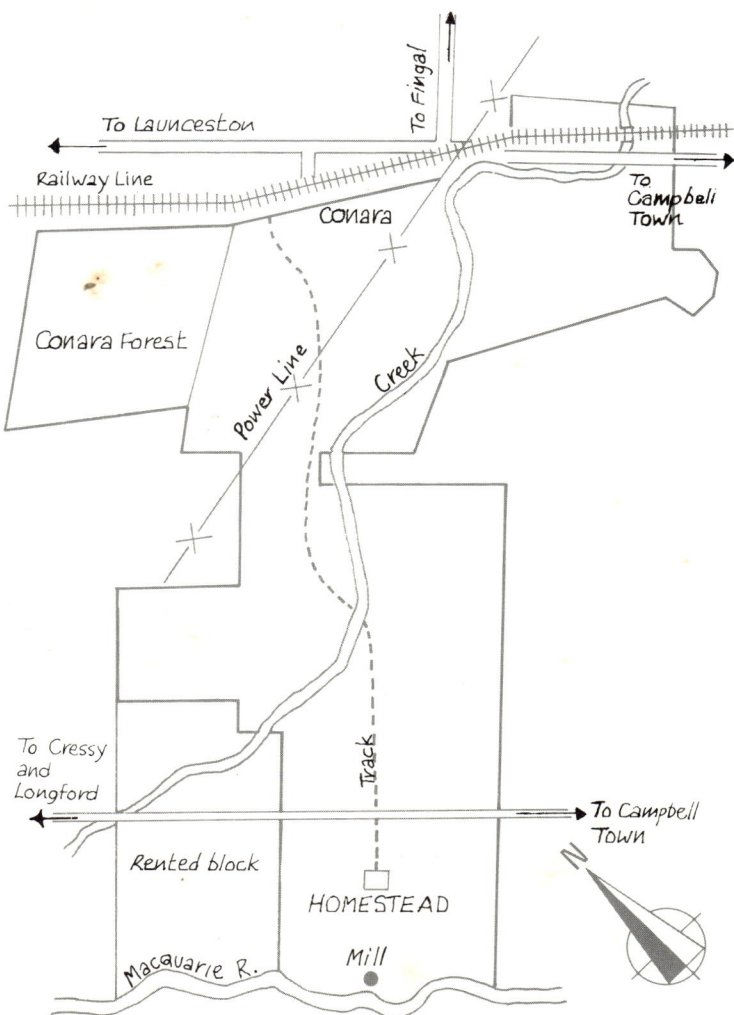

which was bought recently is still untouched bush. On the other side of the river a range of mountains called the Great Western Tiers provides a blue backdrop, whilst in the opposite direction Ben Lomond is massive in the distance. It is quiet, even serene. It is suffering from drought.

Somehow one does not expect a drought in Tasmania. Surely the mountains must trap the rain being carried by the wind from the vast oceans to produce its wild rivers and its abundant temperate growth. Mainland Australia might be dry and withered, but surely Tasmania will be forever green.

But it isn't. The drought, in this part of Tasmania at least, is nearly as bad as that on the mainland. Winton, which would normally receive about 20 inches of rain each year, has so far had only 13 inches. It is the beginning of December and there will not be much more before the end of the year.

The greenness of Winton has long gone. In its place are shades of brown and yellow and even the trees look dusty and lifeless. In the paddocks there is no growth, only the dried out remains from last season. Many of the waterholes have dried up and the creek which crosses the property stopped running weeks ago. The bores are still pumping water into troughs, but it is not enough.

It was in 1822 that George Taylor left his home at Balvaird in Scotland on the long journey to Van Diemen's Land, then a convict colony. On his arrival he was given a grant of land and settled along the Macquarie River on a property he called Valleyfield. In 1830 he exchanged some land with a neighbour called Roderick O'Connor and in the process acquired one of O'Connor's properties on behalf of his son David. It was called Winton.

In 1835 David Taylor and his brother Robert bought some sheep from the redoubtable Mrs Forlonge, who had a neighbouring property. In the 1820s this remarkable Scottish lady had visited Germany with her two sons to buy sheep. Travelling on foot, they had selected Saxon sheep that were pure descendants from those given to the Elector of Saxony by the King of Spain in 1765.

Buying only the best and the purest, the Forlonges drove them slowly to the port of Hamburg, where they shipped them across to Hull on the east coast of England. They then walked them across England to Liverpool and there the Forlonges and their sheep boarded a ship for New South

Wales. But when the ship called at Hobart, Governor Arthur persuaded the Forlonges to settle in Van Diemen's Land instead and they took up land not far from Winton.

The sheep the Taylors bought were direct descendants from this flock and could therefore trace a line right back to those early, pure Saxons. They were the foundation of the Winton stud.

A few years later Mrs Forlonge decided to leave Van Diemen's Land to start a new property in Victoria. The Taylors bought her run and half the remaining flock of Saxons and shortly afterwards acquired some pure Saxon rams that had been imported from Germany by Captain Bell. David Taylor crossed these rams with the Forlonge ewes and the result was a fleece of the very highest quality.

There was already a homestead at Winton which had been built about 1820, before the Taylors acquired the property. Now, David Taylor started to build the woolshed and other buildings which can still be seen today.

Using free labour (unusual in a convict colony), all the bricks were made by hand from clay found on the property. This long, laborious job was made even more difficult when much of the free labour disappeared to join the Victorian gold rushes. But the result was outstandingly successful and these beautiful brick buildings are amongst the best of their type in Australia. Even today new shearers often ask where the woolshed is without realising that they are standing in front of it.

Inside, the walls are covered with notes and drawings made by earlier generations of shearers and station children and there is a fine beamed roof with timbers held together by wooden pins instead of nails. The shingle covering can still be seen from the inside, although the outside of the roof has since been covered with corrugated iron.

David Taylor died in 1860 and after two of his sons had run Winton for a short time it was eventually taken over by another son, John Taylor. In 1890 he built the present homestead, a low rambling building with verandahs on three sides. Its rear makes up one side of the square that contains the earlier buildings.

Still working with sheep descended from the original Winton flock, he bred an outstanding ram called Primus, which in 1896 was declared the Grand Champion of the Sydney Sheep Show. It stamped the Winton flock for years to come.

It was John Taylor who started the tradition of naming the eldest son of each generation John, a tradition which still continues. So when he died in 1919, another John Taylor, his son, took over the property. Five years later he made history when he became the first Australian to export live merinos to Europe. The wheel had turned full circle.

Although he did not retire until 1939, his son, John M. Taylor took over the effective running of Winton in 1936. When war broke out he was amongst the first to volunteer, but to his fury he was discharged and sent back to run Winton. In those patriotic days there were plenty of people

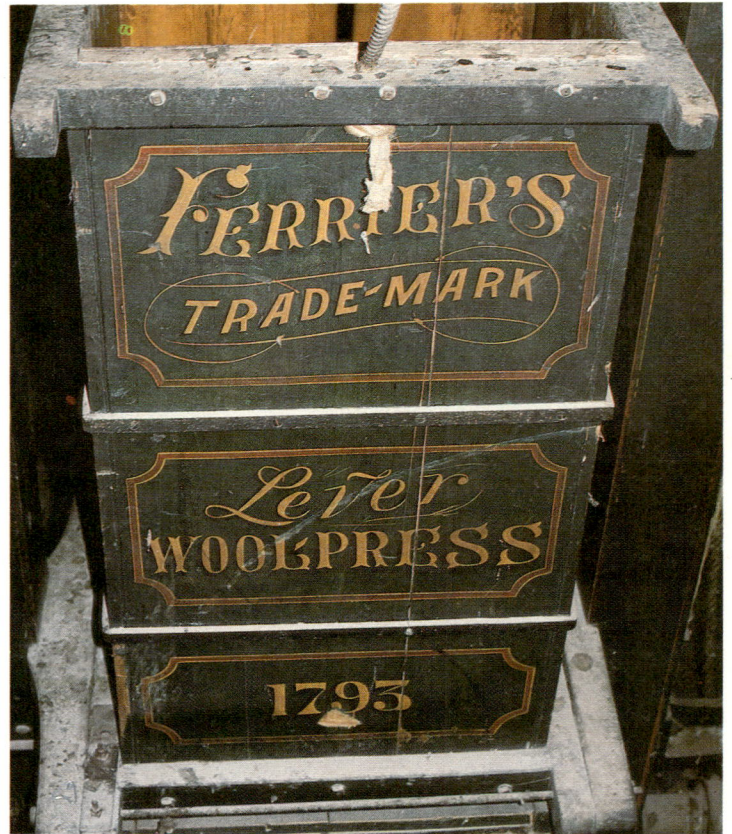

An old wool press which is still kept in the woolshed at Winton.

The homestead at Winton, built in 1890.

Preparing to take a mob of sheep from the yards.

Looking across to the homestead from the door of the woolshed.

An early windmill on the bank of the Macquarie River, with a modern counterpart alongside.

The woolshed, built in the 1840s with bricks made from local clay.

to carry rifles, but very few with the skill to run such an important merino stud.

It was a difficult time. He had six men at the start of the war, but most of these were soon in the forces and he had to manage as best he could. Later he was sent some Italian prisoners of war, but they were not much help. When he discovered that one had been a hairdresser he immediately put him in the shed and gave him a sheep to shear. He didn't give him another.

In recent years his son, another John Taylor, has taken over much of the running of the property. Now thirty-four, he was brought up on Winton and after going to school in Launceston he worked on the property for a year before going as a jackeroo to Sierra Park in Victoria. After two years' national service he returned to take up the reins at Winton.

Although the property itself was in good shape, they were facing major difficulties in the market. Saxon sheep produce an incredibly fine wool not unlike cashmere and it is the highest quality of wool produced in Australia. But in the 1970s demand was very low, partly because of the inroads made by synthetic fibres and partly because the clothes then fashionable did not need a fibre of such quality. The Australian Wool Corporation already had a considerable stock which it had bought in at auction in order to maintain the floor price.

In 1976 John Taylor visited England, Europe, Canada and Japan to investigate for himself the prospects for superfine wool. And it was in Japan that his confidence in the future started to grow. When he returned, many Australian growers were saying that superfine was finished and many in Tasmania had already introduced stronger merinos into their Saxon flocks. But John now thought that the downturn was simply part of the market's cycle and that it was only a matter of time before demand increased again.

In 1978 the International Wool Secretariat, concerned about the huge stocks that had accumulated, launched an aggressive international campaign to promote this wool as Merino Extrafine. It was more successful than anybody had anticipated and within a short time the backlog had been cleared, through sales mostly to Japanese mills. Today, one of them markets its most exclusive cloth under the brand name of Winton.

Whilst the demand for this wool is now even stronger than it was then, many of the problems remain. The Saxon merino might produce the finest wool, but unfortunately it does not produce very much of it. It is also so fine that dust can easily penetrate the fleece whilst it is still on the sheep's back. By the time this dust has been removed at the mill the amount of useable wool is reduced even further.

Whilst superfine has always commanded a higher price than stronger wools, the problem now is that it is not high enough to compensate for the very intensive care the sheep require and the small clip they produce. To keep dust out of the fleece the bare ground near gates should be watered before the sheep are moved through. But with a permanent

David Taylor, who established the Winton stud in 1835. The property has been owned by the same family for nearly 150 years.

The present manager of Winton, John Taylor, examining trophies with his father in 1955.

staff of only two men and a drought as well, this is no longer practical. Even now, the stud rams are still shorn with blades instead of machines. It is an old skill and one that does not come cheap.

John Taylor is built like a footballer and is probably as fit as one. His wife Vera is the daughter of a soldier-settler who took up a small property a few kilometres away and managed to make a go of it. She understands the land and takes an active interest in Winton, although most of her time is taken up with their three children. They live in a comfortable house, just beyond the homestead garden, which they extended a few years ago when times were good.

Winton stud ram JT147, photographed about 1940.

Looking across the drought-stricken paddocks of Winton towards the Great Western Tiers.

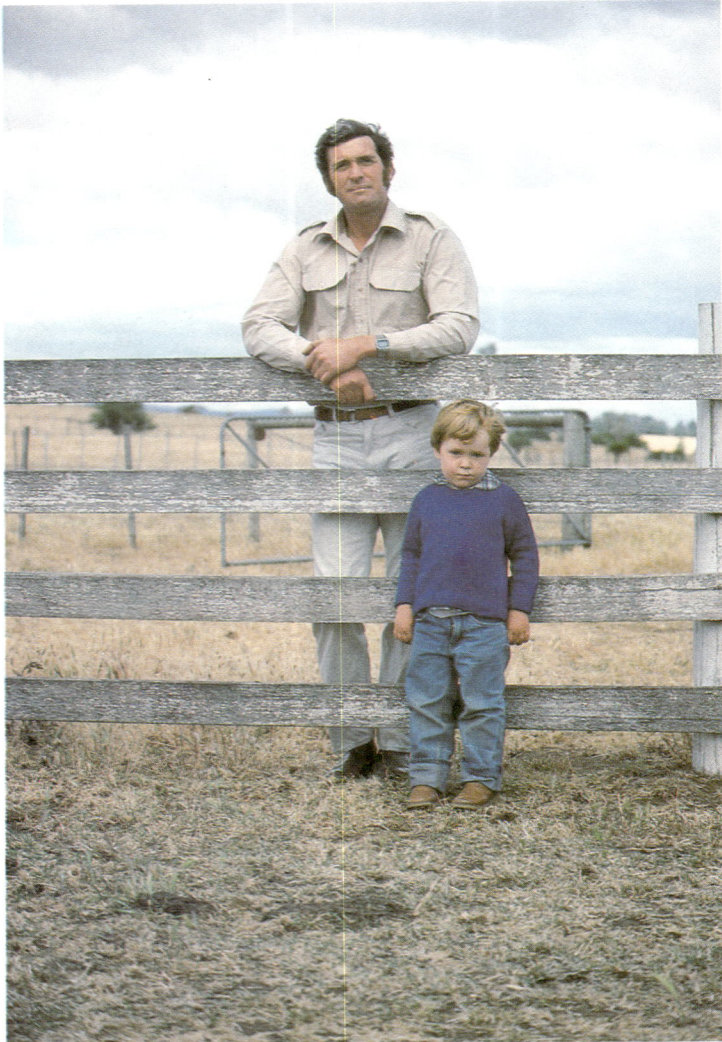

John Taylor, the present manager of Winton, with his son.

Pure Saxon rams from the Winton stud. These rams are direct descendants
from those given to the Elector of Saxony by the King of Spain in 1765.

Moving a mob of sheep at Winton.

The clouds looked promising, but they produced little rain and the drought continued.

Now, times are anything but good. Although they produce outstanding Saxon stud rams demand for them is small
because of the uncertainty about prices and the economic
difficulties of running them. They sell about 60 a year and
produce 115 bales of wool from the flock, but the income
barely covers the costs. One of the children is another John
Taylor. Not yet at school, he should be the next Taylor to
take over at Winton. The question is whether there will be a
Winton to take over.

After nearly 150 years of family ownership, spent running
sheep directly descended from the early Saxons and which
are still absolutely pure, it is a difficult problem. The stud is
almost part of Australia's heritage, but it is also a business on
which people rely for a living. There is a beginning and an
end to most things and after so much time and so much
careful breeding, this drought might be the beginning of the
end.

As well as the 6,000 sheep on Winton there is also a herd
of Hereford cattle which is almost as important. They are
directly descended from cattle bought in 1875, which in turn
came from the first Herefords to arrive in Australia and
which were imported by the Cressy Company in 1820.
Commercially they are very successful, but the drought
means that there is no longer enough feed for them.

In the morning their financial consultant arrives for a
routine meeting. John and his wife sit with him round the
dining table and examine the problem. They have the offer
of agistment in northern Tasmania for 300 cows and calves
at $900 a week. It will cost about $500 a load to get them
there.

On the other hand John can take up a forest lease which is
available to the north-west. The button grass there will
support the cows in their present condition but it will not
support the calves as well and they would have to be sold.
They are still in good condition and should fetch a good
price. Agistment is the better choice, but there is a limit to
how long they could afford it and the difficulty is that
nobody knows how long the drought will last.

In the end they decide to send them on agistment and
hope it does not last too long. If it does, it will cost a lot of
money and in the end they will have to sell the cattle for
whatever they can get. Thousands of dollars are at stake and
there is no way of knowing whether they are doing the right
thing or not. That is what a drought is like.

Later in the day clouds form on the Tiers and blur their
shape. It is raining there and it might reach Winton. The
clouds build up and slowly spread over the river plain. In a
few minutes rain starts to fall. Soon the sky is overcast and
the rain has turned to a light drizzle. Everybody watches,
hoping. Then the cloud breaks up and shafts of warm
sunlight reach down from between the gaps. The sky clears,
the rain stops. And the drought goes on.

BOONOKE

North of Deniliquin in New South Wales the country is so flat that it seems almost unnatural. From the road you can see for miles and if you could find something to stand on you could probably see for ever. There are a few trees around the creeks and rivers, but that is all. On the plain the saltbush, small shrubs like balls of wire netting hung with grey, dusty leaves, stretch all the way to the horizon, and that is all there is. The ground there is as flat as it is at your feet. They say that at dawn, before the sun puts a shimmer in the air, you can see the curvature of the earth.

It is bleak and arid country which you might think would make a good bombing range and little else. But you would be wrong. For this is the Old Man Plain and some of the best merino country in the world. To a sheep man it is what St Andrews is to a golfer, for it was here, on this seemingly inhospitable country, that the Australian merino first came into its own.

Men had introduced the merino to Australia long before this Riverina country was settled. Some have since been credited with founding the Australian wool industry, but they have been given too much credit. For it was here, on the Old Man Plain, that the Australian wool industry really began.

In December 1850 George Peppin and his two sons, George and Frederick, arrived at Port Phillip Bay. They had come from Somerset in England where they had kept merinos which were descended from the King's flock. Ten months after they arrived they bought the lease of a property near Mansfield in Victoria called Mimaluke, later to be covered by the waters of Lake Eildon.

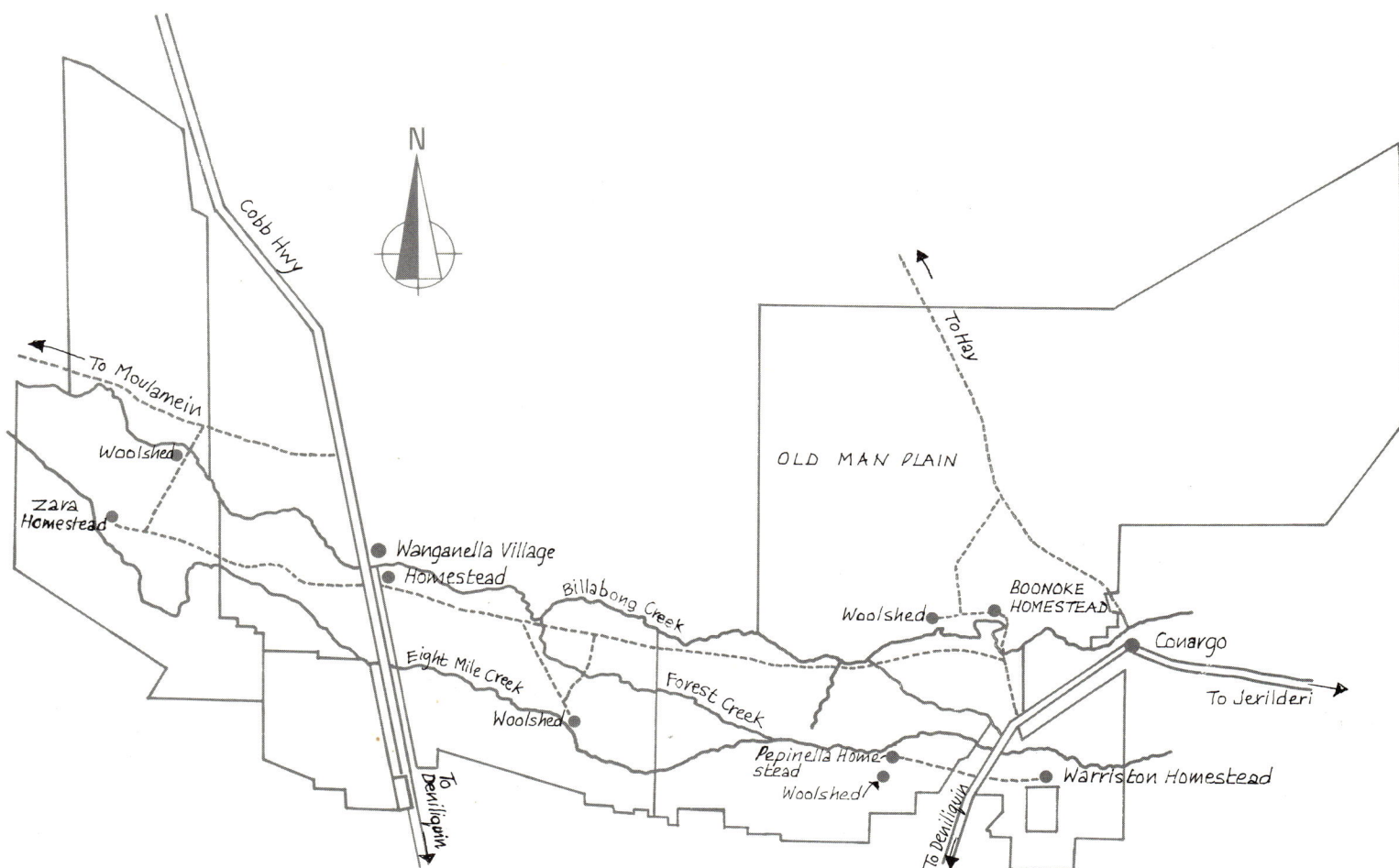

N

Cobb Hwy

To Hay

To Moulamein

Woolshed

OLD MAN PLAIN

Zara Homestead

Wanganella Village

Homestead

Billabong Creek

BOONOKE HOMESTEAD

Woolshed

Conargo

Eight Mile Creek

Forest Creek

To Jerilderi

Woolshed

Pepinella Homestead

Warriston Homestead

Woolshed

To Deniliquin

To Deniliquin

In 1858 they sold out and paid the considerable sum of £10,000 for a property called South Wanganella on the Old Man Plain. It consisted of about 52,000 acres and the price included the stock of about 8,000 sheep. To this they added their own stock which they overlanded from Mimaluke.

It was not a successful move. Although they bought the neighbouring property of Morago in 1859, the following year the whole of Wanganella Station was up for sale. But when no offers came the Peppins realised that they would have to carry on as best they could. In which case, they thought, they had better start improving the place.

Their sheep at that time were no better than those elsewhere and they decided that this was where the first improvement should be made. They would try to breed a vigorous, medium-wool sheep and they called in Thomas Shaw to help them do it.

Thomas Shaw had very strong views about Australian sheep. Whilst still in England he had written of the wool buyers' concern about the poor quality of wool imported from Australia:

There was a mystery about the wools of these colonies which they could not unravel. . . They were not breeders of sheep, but judges of wool; thus they could not comprehend how it was that wool from various flocks . . . should so suddenly lose many of the qualities that it had been years in gaining. . . The wool was Australian, retaining some of its leading qualities; but it was Australian wool spoiled.

When Shaw came to Australia he soon found out why. The breeders had, he said, developed a kind of composite breed of their own, a breed that was

in their estimation the perfection of human wisdom — but in reality a mongrel breed, in which may be found every shade between the real Australian merino and the dried-up Leicester, mixed with myriads not fit to class as respectable goats.

He drew hostility from all directions, but he was just the man for the Peppins.

Shaw selected 200 ewes from the Peppin flock and bought another hundred from a station at Balranald. They then crossed these with selected Peppin rams and others imported from nearby properties. Ahead of their time, they were trying to develop a type of wool that

the country would grow, instead of trying to produce what the climate and soil would continually fight against.

And they succeeded. By the time George Peppin died in 1872 they had developed a merino that was ideally suited to the conditions found in much of Australia. This was to make his name famous amongst sheep men ever since, for the animal he developed was to become the corner stone of the Australian wool industry — the Peppin merino.

The two sons took over after his death. Helped by good sales of wool and sheep, in 1873 they bought the neighbouring property of Boonoke and stocked it with Peppins from Wanganella. By the time George died three years later they were enjoying considerable success, but his brother Frederick

A Boonoke stud ram.

Boonoke keeps a contract shearing team fully employed for about six months every year.

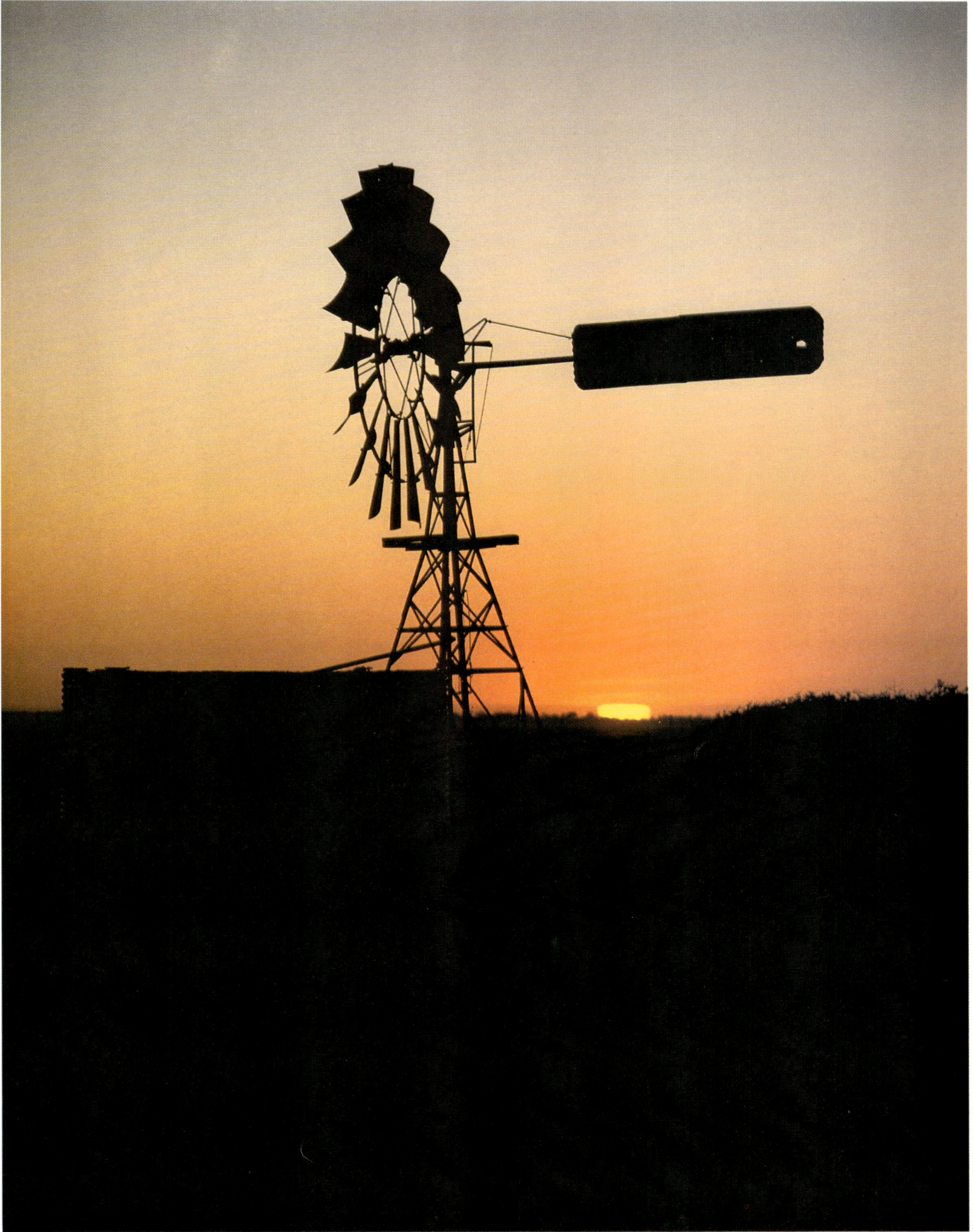

Sunrise on the Old Man Plain.

had taken little part in the business and he decided to sell the property and concentrate on interests elsewhere.

The property now consisted of 172,441 acres of freehold and Crown lease land and about 55,000 sheep which included the famous stud flock. It was much more successful than when it had first been offered for sale but, surprisingly, buyers were just as scarce. When it came up for auction at the Menzies Hotel in Melbourne there was not one bid. But a few weeks later, in November 1878, it was sold privately to Albert Austin and Thomas Millear, who paid £70,000 for Wanganella and part of Boonoke, together with the stud flock and 28,000 sheep. They had been offered the whole of Boonoke but Millear was already over committed and they had to refuse it. Instead, it was sold a few days later to F. S. Falkiner.

Franc Sadleir Falkiner was born in Ireland and had arrived in Australia in 1853 to look for gold. Then only nineteen, he opened a store at Maryborough and three years later married Emily Bazley, who was then only fifteen. After a few years as a successful trader he turned his attention to land and moved to Geelong as a government land valuer. However when one of his sons fell ill, his doctor thought the hot, dry climate of the Riverina would be more suitable. So, with two partners, Franc Sadleir bought the remainder of Boonoke and 26,000 Peppin sheep for £60,000.

Whilst Austin and Millear continued to develop Wanganella, Falkiner did the same at Boonoke and four years later he was able to buy out his partners. But with his forceful personality turbulence was never far away. In 1888 his friend and neighbour Frederick Parker accused him of cheating at the Deniliquin Show by entering two rams in the grass fed section that had been specially fed. Falkiner was barred for life and Parker sued him for slander after Falkiner said that Parker had himself done the same thing the previous year.

A jury in Deniliquin heard the case and, against all the evidence, found in favour of Falkiner. Parker then complained that Falkiner had secretly knobbled the jury and a new hearing was ordered. Falkiner finally settled with Parker before the case came into court and in the process probably established a world record price for supplementary feed for two rams.

Boonoke and Falkiner continued to prosper and he bought more land as the flock grew. In 1884 he bought Moonbria Station about 30 km from Boonoke and sent his eldest son, seventeen-year-old Bert Falkiner, to run it.

Four years later, whilst in the grip of drought, the homestead at Boonoke was completely destroyed in a fire thought to have been started by a disgruntled shearer who was trying to establish the union in the woolshed. The present homestead was built two years later from a set of English plans that had already been used on Tuppal, a neighbouring property. After the drought ended in 1891 Falkiner bought Tuppal. In only thirteen years his original 30,000 acres had expanded to a massive holding of more than 250,000 acres, and during that time he had sold nothing but sheep and wool.

Frederick Peppin.

Franc Sadleir Falkiner, who bought Boonoke from the Peppins in 1878. By the time he died in 1909 he owned half a million acres and was known as the Napoleon of the Old Man Plain.

28

Otway Falkiner, the second Napoleon of the Old Man Plain, who ran Boonoke from 1915 until his death in 1961.

An early photograph of the Boonoke homestead. It was built in 1890 from a set of plans which had already been used by a neighbouring property.

In 1894 the rabbits reached the Riverina and in that year alone Falkiner spent some £120,000 in fighting them. Rabbit-proof fences were built and rabbits poisoned by the thousand. One of his sons wrote:

One of the worst jobs I had to do was to clean out the rabbit bays along the fences. Dead and stinking rabbits were waist-high around me as I worked.

Unlike many graziers, Falkiner had the capital necessary to beat off the plague and he did not hesitate to use it.

By now Falkiner was an old man. In 1898, worried that the stud might be broken up on his death, he formed the company that still trades at Boonoke — F. S. Falkiner & Sons Limited. He gave shares to each of his sons, but not to any his five daughters.

Meanwhile he continued to buy more land and by 1901 the property consisted of six stations totalling nearly 405,000 acres. That year they shore 394,000 sheep, produced 6,589 bales of wool and sold over 4,000 rams. It was grazing on an enormous scale. But the following year they were struck by one of the worst droughts on record and prosperity turned to survival as ram sales slumped by half and sheep died in empty paddocks.

All the living sheep were concentrated inside the banks of a dam which was nearly dry. The only water in it was shallow and appeared to be covered with scum. For some distance from the water's edge were the bodies of dead and dying sheep which had been bogged in the mud. Others had drowned in the muddy water. Those which could still walk were staggering over the bodies only to fall into the water when they reached it.

Dreadful though it was, the Falkiners of Boonoke survived and in 1904 the company reported a profit of £64,000. But Falkiner refused to pay a dividend, saying that liabilities had to be reduced first. He relented the following year and paid the princely dividend of 1 per cent. This was repeated the following year, even though the profit was then nearly £220,000 and ram sales alone had produced more than £40,000.

When Falkiner, now known as the Napoleon of the Old Man Plain, died in 1909 the property consisted of more than half a million acres carrying a quarter of a million sheep, seven thousand head of cattle and nearly a thousand horses. He had come to Australia a poor man and succeeded where many had failed.

After his death things changed quickly and radically. Bert Falkiner took over the company. The following year it paid a dividend of 14 per cent. The profits were certainly high, but most came from land which had been sold because of the prospect of a Labor State government and the fear of a new land tax that it would introduce. But when Millear offered to sell them part of Wanganella they could not resist it and they paid £125,000 for 2,400 acres and the stud flock of over 100,000 sheep.

By now the Falkiner sons were mature men and many had developed interests elsewhere. Bert Falkiner resigned from the company in 1915 and moved to Sydney to look for a

Going to work — complete with dogs.

property of his own. Eventually he found one that had been stocked with rams from Wanganella. On the north-west plains beyond Dubbo, it was called Haddon Rig.

Boonoke, now run by Otway Falkiner, continued to flourish. As a result of more land purchases, by 1917 it was the biggest and most famous merino stud in the world, with sales of stud sheep alone producing more than £60,000 each year.

This prosperity continued until 1929, when the Commonwealth government banned the export of live merinos. The property went into decline with the Depression that followed and in 1932, with stud sales producing only £23,000, some 50,000 acres of Boonoke were sold.

Otway Falkiner started to restore the property with the development of a new flock. He had noticed that some sheep did not develop horns and that these were easier to handle and were less likely to get caught in fences. He selected the best and from them bred the first flock of poll merinos.

But there were soon to be more setbacks. In 1945 they experienced the worst drought since 1902 and although some £98,000 was spent trying to keep sheep alive, they lost nearly 10,000 before the drought ended. It was a testing time for rural properties throughout most of Australia, but Otway Falkiner never lost his confidence in the excellence of Boonoke sheep.

Once, when one of his station hands was charged with an offence against a sheep, he agreed to give him a character reference at the magistrate's court in Deniliquin. He started, as everybody knew he would, by describing the excellence of the sheep he bred and how their quality was world famous.

'Yes, yes, Mr Falkiner', said the magistrate testily after ten minutes. 'We all know that and have heard it many times. But you are here today to speak on behalf of this man.'

'Oh, that,' said Falkiner. 'Well, I give each of my station hands two sheep a month. They can either eat them or stuff them and I don't care which.'

Slowly the market improved until the next drought in 1957. Otway Falkiner, now eighty-two years old, held a huge sale of stock at Boonoke which was attended by over four thousand people. The buyers, most of whom were also affected by the drought, were hesitant. But by the end of the day they had spent over £63,000 and much of it had been paid for the new Poll Boonoke ewes which had been offered for the first time.

Later that year Otway Falkiner suffered a stroke that was to keep him in a wheel chair for the rest of his life. But it could not crush this giant of a man. Whilst at a local race meeting he was taking a pull of scotch from a friend's hip flask when his nurse returned and tongue-lashed both of them.

'What's the matter?' said the friend. 'Are you frightened he'll wet his bed tonight?'

'No,' she said. 'I'm frightened he'll wet mine.'

Alec Morrison, the Executive Director in charge of Boonoke.

Henry O'Connor, in charge of the Stud Division.

Peter Hansen, Supervisor.

Bob Sefton, General Manager of Boonoke.

The Boonoke homestead today.

By the time Otway Falkiner, the second Napoleon of the Old Man Plain, died in 1961 he had had the satisfaction of buying the remaining part of Wanganella, uniting the two properties for the first time since the Peppins. But it was the end of an era, for there was no third Napoleon waiting to take over.

Wool prices fell dramatically and after a number of losses the family announced what business men in Melbourne had suspected for some time: the company was for sale. The asking price was $4 million, but with wool prices still depressed there was no rush of buyers. Later that year it was sold for $2.6 million to Frederick James, who in 1949 had started the Cleckheaton spinning mills.

As wool prices climbed, prosperity slowly returned to Boonoke, but when James died his company decided it should concentrate on its mills and once again it became known that Boonoke, or at least part of it, could be bought. With the property again in drought, a half share was available for $3 million. Again there was no rush, but in 1978 an unexpected visitor arrived by air to look it over. It was newspaper publisher Rupert Murdoch who, after a quick inspection, returned to Sydney for lunch.

In spite of his apparent lack of interest, Murdoch was thinking very hard about the deal. He had been brought up on the land and had a property of his own near Yass in New South Wales. He was well aware of the importance of the Boonoke stud and after a few months he decided to buy it. But he was not interested in a half share — it had to be all or nothing. In the end he bought the whole of the property for

the remarkable price of $3.5 million — remarkable because the stock and equipment alone were probably worth that much. A few weeks after the deal had gone through, rain fell and the drought was broken.

Any idea that Murdoch had bought the biggest hobby farm in Australia was soon dispelled. It was a business, and it would be run like one. Soon millions of dollars were being spent on irrigation to protect the property from the droughts that had been such a recurring feature of its history. At the same time the staff were given very clear instructions to develop the very best sheep, sheep which could be sold with pride to clients who had always relied on the Boonoke name for quality and performance.

Today, Boonoke is a huge sheep station of some 228,000 acres. It consists of five adjoining properties, including the original Wanganella Station, each with their own homesteads. There is also another property of 5,000 acres in Western Australia.

Boonoke employs about forty people, although with families there are about a hundred people living there. Of these forty, ten or twelve are jackeroos, eight are managers or supervisors, and the rest are station hands. It seems a small number for such a vast area, running some 90,000 sheep. Indeed until the 1960s there were about 150 people doing virtually the same amount of work. Two things have brought about the change: motor bikes and efficient management.

With huge paddocks many kilometres from the homestead area, the value of motor bikes is soon apparent. A man can go to the furthest paddock in the morning, do a day's work,

The old ram shed at Peppinella.

and still be back at the end of the day, with him a detailed account of what he has done and what he has seen on the way. With a horse the same job might have taken days and would probably have needed more than one man as well. Rations would have had to be carried, the horses looked after, shelter provided — all to do what is now a day's work for one man.

Even the dogs don't get left behind. At the sound of a whistle, often given as the bike moves off, they leap on to the pillion seat and somehow manage to stay there as the bike roars off in a cloud of dust. If they are not whistled up, they go miserably away to find some shade under the trees and wait for next time.

Efficient management, though less noticeable, is just as important. The whole property is run from one modest office building. The considerable amount of paperwork generated by a business of this size is handled by one full-time accountant and a typist who comes in two days a week. Indeed, for much of the day the offices might be empty and when the managers are there it is because they are pinned down by some essential routine when they would much prefer to be out in the paddocks. And given half a chance

they will find an excuse to escape, for these are active men who do not take kindly to a desk.

The Executive Director in charge of Boonoke, Alec Morrison, is no exception. Born in Dubbo in 1935, he was educated at Scots College in Sydney and there won a Commonwealth Scholarship to study law. But he knew by then that that was not the life for him and in 1952 he went as a jackeroo to Terrick Terrick in Queensland. He joined the Falkiners as a jackeroo in 1955, together with Bob Sefton who is now general manager of Boonoke. He later became manager of another Falkiner property and when that was sold in 1958 he joined Elders in Sydney as Stud Stock Officer for New South Wales. In 1961 he left to set up as a freelance sheep classer at Dubbo and was soon classing the Wanganella flock for the Falkiners.

When Fred James took over Boonoke he asked Alec Morrison to join him as marketing manager. He did so, attracted by James's enthusiasm at a time when most people in the industry could see nothing but gloom. He stayed on when Murdoch took over, but not without reservations. He was reluctant to go through another change of ownership but decided to stay in the hope that the new owner would bring about the changes and improvements that were clearly needed. He was not disappointed.

In the past each station had been run as a separate property under the control of a manager who was responsible for that property and no more. It was government by geography, with the result that much of the work was duplicated on each station.

Today, Boonoke is organised into divisions which are based on activities rather than on geography. There is a maintenance division which is responsible for all the hardware, and a commercial division which is run on Zara, a property bought in 1927. Both these divisions are run by Bob Sefton.

Then there is the important stud division on which much of the reputation of Boonoke rests. It is under the control of twenty-five-year-old bachelor Henry O'Connor, who lives alone in the old homestead at Peppinella. Tall, lean and quietly spoken, he handles this vitally important job with a confidence that belies his age. Totally at ease with himself in spite of his responsibility, he says simply that Boonoke is his paradise. And in spite of the seemingly unwelcome nature of the country, Boonoke is something of a paradise.

With a rainfall of only 14½ inches a year, it depends heavily on the permanent water of Billabong Creek which runs east to west across the huge area of Boonoke. The untrained eye is grateful for the splash of green provided by the willows along its banks, but the expert knows the value of the creek is far greater than that: it is the very life blood of Boonoke. It not only provides water for the sheep who crowd its banks, but also to irrigate the paddocks themselves.

Irrigation is an expensive business. The paddock is graded with a laser to an accuracy that would seem impossible with such a big area. When it is finished the paddock looks completely flat, but it isn't. It has a minute slope on it so that when water is released from a channel at the top of the paddock it runs slowly down to the other end. With good judgement there will be hardly any surplus water in the drain at the bottom of the slope and the whole paddock will have been thoroughly watered with little waste.

Wheat is grown to stabilise the soil and then the paddock is put to its normal use as grazing for the sheep or for growing millet for feed.

Out on the plain there is no irrigation, nor any sign of surface water. But, far apart, windmills raise water from below the ground and pour it jerkily into the tanks alongside. From there it is fed into a series of troughs to provide plenty in the middle of nothing. At night, when the wind drops, the windmills stop turning but in the early morning, as the sun rims the horizon, the breeze springs up again and, with a muted rhythmic clanking, the windmills start to turn again. It is like a heartbeat starting, to give life where it would not exist.

In the homestead area huge pepper trees provide shade in what looks like a village square. In one corner is the white-painted office building and alongside it is a tall mast which provides radio communication between the office and

Relaxing on the homestead verandah at the end of the day.

Now parked beside the drive, this old dray is a familiar sight to Boonoke visitors.

the vehicles in the paddocks, often many kilometres away. On one side of the square is the jackeroos' quarters and on the other side is the back of the homestead. The front, with its elegant iron-work verandah, looks across acres of well-kept garden. Inside the tall and spacious rooms open out from lengthy corridors, dark and cool against the afternoon sun.

The ram shed stands on higher ground and is surrounded by lush green pens that stand out vividly against the surrounding country. Here live the Boonoke stud rams, the aristocrats who represent the highest art of Boonoke stud breeding. In 1980 a half share of one of them, Woolmaker 80, was sold for $20,000 and already it looks as if it was a bargain. Proud looking, these magnificent animals seem to know that they are the elite of Boonoke.

Important though these expensive stud rams are, the real business of Boonoke is to provide good quality sheep to commercial growers who might buy ten or twenty sheep each year and who expect to pay realistic prices for them. Because of this, the company recently decided to withdraw from

competitive showing. This was a difficult decision, especially as decades of successful showing had brought ribbons and championships almost beyond count, but it was thought that these shows no longer had much relevance for the commercial buyer.

Instead, Boonoke now holds annual Field Days on the property so that the buyer can see hundreds of typical Boonoke sheep, instead of the handful of specially prepared animals he would see at a show. And, equally important, he can see the type of country that has produced them. These Field Days are now an important part of the Boonoke year and people come from all over the country to look at sheep and to talk to the experts who have produced them.

They sell about 6,000 rams each year at Boonoke, more than any other Australian stud. In addition they produce about 2,500 bales of wool a year and keep a contract team of shearers from Deniliquin fully employed for half the year.

The men of Boonoke are proud of its history and the role the property has played throughout its long past. But it is a young team and its concern is with the future. Their jobs

might be far removed from those of managers in high-rise offices of Australian cities, but their approach is the same. Money has to be put to good use, targets have to be set and met, cash flows have to be prepared and customers have to be looked after.

But unlike managers in the city, they can jump into the ute, whistle a dog up on to the back, and go and have a look. Bouncing across the paddock in search of sheep (for in paddocks this size they can be anywhere), they are men in an environment they know and like. And when they find the sheep they look at them with an unhurried gaze, assessing their condition, evaluating their feed, looking for any in need of attention. If there were fewer of them they would know their names.

It is 7.45 on a sunny morning. Insects are buzzing and the early sun throws long shadows under the pepper trees. It is going to be a hot day and the dogs are already lying in the shade, waiting for work to begin. Utes pull up in front of the office and men go inside as the dogs in the back greet those under the trees. Soon these men come out of the building, jump into the cab, and drive off again in a cloud of dust which takes a long time to settle.

Children appear, scrubbed pink and looking out of place in school uniforms. At 8 o'clock a small bus arrives and they pile on board, jostling and shouting like school children everywhere. Another swirl of dust and it is gone, taking them to a different world outside. More men come and go, others walk unhurriedly across the square whilst a jackeroo dashes across more quickly because he is late.

It is the start of another day at Boonoke. A day which, like all the others, will be spent trying to produce some of the best sheep in the world. A day not very different to those spent by generations of Falkiners, or the Peppins before them when they were breeding the sheep that founded the Australian wool industry.

In the paddocks the Peppin merinos finish grazing and settle down in the shade to watch the sun climb higher and higher. And out on the Old Man Plain the windmills are turning, as they have for years.

SHEARING

It was once said that if the sheep station represented Australian autocracy, then the shearing shed was a true republic. And to some extent this is still true, for shearers and the shearing have not changed much over the years. It is still hard, physical work and the men who do it are as strong and as independent as those who, in songs and stories, became part of the Australian legend.

Today, most shearing is done under contract and shearers no longer write to stations to book their stands. Instead, the station engages a contractor to supply a team of shearers and other staff for the duration of the job. The station pays the contractor, and the contractor pays his men.

Nor do they always live on the station during the shearing. In parts of the country where there are a lot of sheep properties a contractor can often keep his teams employed within a short drive of their homes. Elsewhere, however, the team will still live in the quarters provided for them near the shed. The contractor will then supply a cook, but his wages and the cost of food will be paid for by the members of the team.

Shearing starts each day at 7.30 in the morning and is done in four 'runs' of two hours each. They are divided by morning and afternoon smokos, each of thirty minutes, and a lunch break of one hour.

In the shed the shearers work at a series of individual stands that run the length of the 'board'. At the start of the run, each shearer will take a sheep from the catching pen and drag it on its haunches to his stand. He will put the sheep in position, take hold of the handpiece which is used to cut the wool, and turn it on by pulling a cord that moves a small machine on to a friction wheel on a revolving shaft that runs above his head.

Shearing is done in a sequence of moves and holds that have been designed to remove the fleece neatly and quickly, and with no more effort than absolutely necessary. Bending almost double, the shearer starts with a series of cuts down the belly towards the groin so that this 'belly wool' comes away in one piece and drops as a small tangle of dirty wool on to the floor. This is picked up by a rouseabout and thrown into a large three-sided cubicle, called a bin, which has been set aside for these pieces.

Meanwhile, the shearer works up the inside of each back leg and continues up the outside of the flank furthest from his cutting hand to work over the rump. He then leaves this part of the fleece and cuts the wool on the top of the head between the ears, called the topknot. He then moves the sheep so he can hold its chin up and starts to cut upwards from where he made the first cut down the belly. Working upwards now, he works round the neck, which contains the thickest part of the fleece, then round the face and across the back of the head to join up with the previous cut between the ears.

He now opens up the wool across one side of the ribs from the belly then, with the sheep on its back, he makes the 'long blow', the first of a series of sweeps that

The shearer removes the fleece with a sequence of moves that are quick and efficient and which use no more effort than absolutely necessary.

Using a wool press to pack wool into the bale.

run from the rump right up to the back of the head and which extend across to the backbone. He then works round the ear and down the neck to the top of the shoulder on the other side of the sheep and follows this with a series of cuts from the spine round the far ribs to the belly. He works down the flank, called the 'whipping side', with a series of short cuts to the hind leg and the fleece finally comes away in one piece, the wool held together by its natural crimp.

As soon as the fleece is released it is picked up by a rouseabout and carried over to a long, slatted table. He throws the fleece over it and as it settles other men work quickly round it, pulling off the stained pieces which they throw into another bin. The wool classer then examines the fleece more carefully to assess its grade and then it is bundled up and placed in another bin that contains other fleeces of the same grade.

Although rare until recently, women are now common in shearing sheds and work as rouseabouts or wool classers.

Meanwhile, the shearer has released the now naked sheep through a trap door that leads to a counting-out pen outside the shed and has taken another from the catching pen. He shares this pen with the shearer next to him and etiquette requires him to select one of the sheep nearest to him in the pen. Some sheep are easier

Throwing the newly shorn fleece on to the classing table.

to shear than others, but he is not allowed to search them out and instead must take the first sheep he touches. An experienced shearer, however, can tell at a glance which it should be.

Whilst he is fetching his next sheep, a rouseabout will be sweeping the loose locks from the board and another will be doing the same round the classing table. These locks are kept in another bin, for they will amount to a good deal of wool by the time the shearing is over.

Behind the bins another man takes the fleeces and feeds them into a machine that compresses them into an empty bale. When it is full, the bale is fastened and then weighed. Using a stencil, he paints the name of the property on the bale, together with a description of the wool, its grade and weight, the registered number of the wool classer, and the number of the bale. The bale is then moved to a storage area to await shipment.

There is a hierarchy in a shearing shed that is not always obvious but which is very important to those working there.

First, the shearers. A shearer who can consistently shear more than two hundred sheep in a day is called a 'gun' and he carries that title wherever he might work, although he might not always work as fast as that. The shearer who shears most sheep in a day is called the 'ringer' and he carries the title in that shed until somebody beats him. Shearers usually draw lots before the start of shearing to decide which stand they will take, but a learner, who will cut fewer fleeces than the others, will be given the stand furthest from the classing table.

The shearers elect their own spokesman, called the 'rep', whose job it is to discuss any grievances on their behalf with the 'boss of the shed'. This will be the contractor if he is present, but if not it will be either the wool classer or the 'expert', the man responsible for maintaining the machinery and for sharpening the cutters used in the handpieces.

The owner of the property can deal only with the boss of the shed. If, for example, he thinks that a shearer is working clumsily and damaging sheep, then he can make his complaint only to the boss, not to the shearer. The boss will then take up the matter with the shearer and convey any repeatable comment he makes in reply.

Shearers will not work on sheep that are wet because of a long-standing belief that wet sheep are dangerous to their health. This remains unproven (although still possible), but as wet fleeces cannot be pressed into the bales because of a danger of spontaneous combustion there is usually little point in continuing anyway.

The shearers alone decide whether the sheep are too wet to work on. Having shorn at least two sheep in the run, any shearer can claim that the sheep are wet. The rest of the team will each shear two more sheep and the

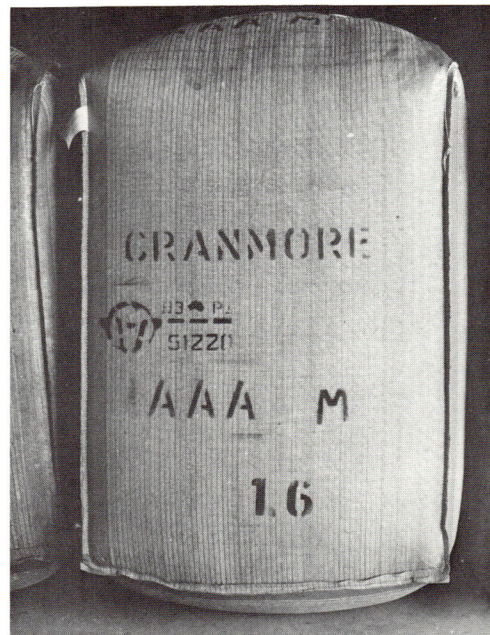

A bale of wool marked with the name of the property, the grade of wool and its weight, the number of the wool classer, and the number of the bale.

matter will then be voted on by ballot. If the majority say they are wet, the rep informs the boss and the men stop work for the day. If they decide the sheep are dry the work will continue but any shearer who has voted 'wet' can stop work if he wishes, although he will not be paid. The station owner and the boss can take no part other than to witness the counting of the votes.

It follows that the station staff must do all they can to keep the sheep dry. Sheep have to be kept in the shed for four hours before shearing so that they are 'empty', but that is not long enough for them to dry out if they are really wet.

A team of five shearers will handle some 750 sheep in a day and it is the job of the station hands to maintain a constant supply. They do this by continually refilling the catching pens from a much larger pen, called the 'sweating pen', that might occupy half the shed, and then refilling that from sheep kept in the yards outside. Other station hands, meanwhile, will be bringing mobs in from the paddocks to refill the yards.

For the owner of the station, shearing is one of the biggest costs of the year. The rouseabouts, expert, wool classer and the presser are paid on a daily rate, whereas the shearers are paid by the number of sheep they shear. The present rate is $93.78 per hundred sheep, which means a shearer will usually earn about $150 for a day's shearing. But he receives no holiday pay or superannuation and if he is too ill to work, or if he cannot shear because the sheep are wet, he will earn nothing.

In spite of this, however, shearers are well paid and always have been. But as they say, if you want that sort of money and can put up with the hard work, there is nothing to stop you joining them.

SPRINGFIELD

About 17 kilometres from Goulburn, along the Braidwood road, is a large sign beside a gate. It says 'Fonthill Sheep Measure Up.' It seems at first to be a harmless attempt to advertise the sheep that are bred at Springfield. But it is much more than that. Measurement is important at Springfield and that sign has more than a touch of defiance in it. Jim Maple-Brown has strong views about measuring sheep and not everybody agrees with them. But as one of the leading innovators of the day, he doesn't expect them to.

The drive from the gate to the homestead is only a couple of kilometres long, although if you take the wrong turn you will miss the homestead completely and go another kilometre or so to a separate group of houses called Pinea. It is like a suburb, but it is all part of Springfield.

The property consists of 7,500 acres and it is divided into about 25 paddocks each averaging about 300 acres. It is pleasant, undulating country and supports about 14,000 sheep. Lightly timbered, it receives annually 25 inches of rain which usually falls evenly throughout the year. The Mulwarree Chain of Ponds, through the middle of the property, provides permanent waterholes even in the worst seasons. The remainder is watered by a number of natural springs and a few bores. Clay tanks store run-off water when there is any. Most of them are empty now and the country is the colour of cornflakes.

Near the homestead is an old mill, a coach house and other farm buildings grouped around an open space. A complete street of small one-storeyed cottages runs from one corner. They would not look out of place in Paddington or Carlton, except that here they have a much better view, looking across the paddocks of Springfield to the distant hills.

The street is deserted during the day. Springfield is worked by a manager, three men and a jackeroo. The jackeroo lives in the homestead and the men live with their families in the cottages. Some of the cottages are also let to people who do not work on Springfield but who pay a low rent in return for maintaining them. It is a good system, for owning a street of cottages in the middle of a sheep station can be an expensive business.

Surprisingly, perhaps, none of the men who work at Springfield were born on the land. The jackeroo comes from Canberra and all the others come from cities too. One of them worked in the steel works at Port Kembla but left when

they kept telling him to slow down. Now he can work as fast as he likes and nothing would get him back.

At first there seems little that is new about Springfield. The buildings are old and pleasant-looking and have the timelesness of an English village. It is solid, comfortable and traditional, but it is deceptive. And that sign at the gate is the clue.

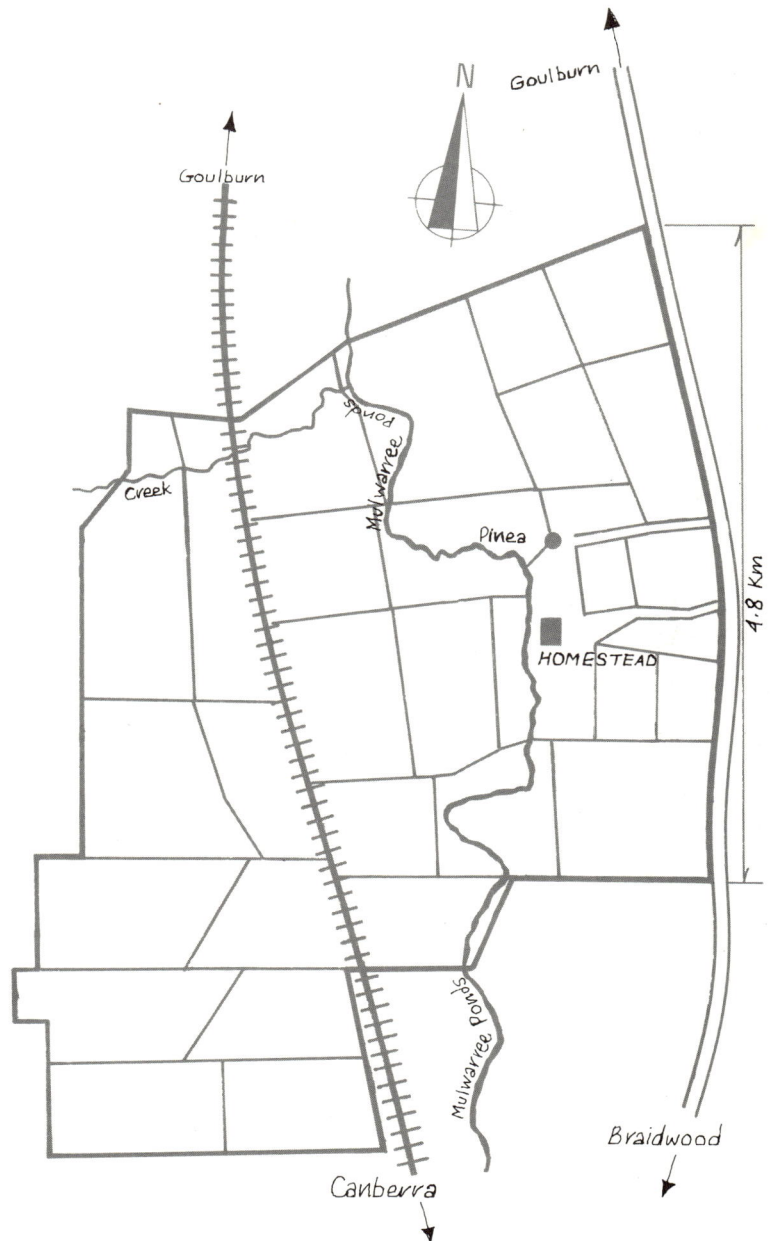

William Pitt Faithfull, who founded Springfield in 1827.

Lucian Faithfull with two of his daughters: Hazel, standing, and Bobbie.

Jim Maple-Brown believes that change is one of the few means of survival, and if that means changing the way sheep are bred and the way they look, so be it. He is something of a maverick in an industry which, until recently at least, has regarded change as being worse than a drought and nearly as bad as warm beer.

Tall and straight in middle age, he still believes in it as passionately as he ever did. He sums it up in a quote from *The Mighty Micro* by Christopher Evans which he thinks expresses it perfectly.

Intelligence is the ability of a system to adjust to a changing world, and the more capable of adjusting — the more versatile its adjusting power — the more intelligent it is.

So if the property is old, the ideas are not.

William Faithfull came to Australia in 1791 as a private in the New South Wales Corps. Like many soldiers in that Corps, he soon found there were more opportunities in New South Wales than there ever had been in Britain. In 1808 he was given a grant of 1,000 acres and he took it up in what are now the Sydney suburbs of Burwood and Enfield. It was not a good choice, however, and a few years later he changed it for better land on the Hawkesbury River near Windsor.

His eldest son, William Pitt Faithfull, was twenty-one when he moved into the Goulburn district to take up land of his own. In 1827 he established himself on a property which he called Gooranganennoe, but the name was so unwieldy that he soon changed it to Springfield.

He proved to be an excellent sheep man and, unlike many at that time, was keen to improve the quality of his flock. In 1838 he bought ten rams from the Camden flock of John Macarthur. These sheep were direct descendants from those Macarthur had bought from Captain Waterhouse and the 'Argot lot' which Macarthur had imported from England in 1804. At about the same time he also bought some pure Saxon ewes from Winton in Tasmania. As a result he had some of the finest breeding stock then available and as he made a point of classing his young sheep (an uncommon technique then), he soon developed a flock that was clearly ahead of its time.

If he was good at producing sheep, he was just as good at rearing sons, of whom he had several. On the morning of Monday 6 February 1865 four of them climbed into the four-in-hand to set off for Goulburn. The two youngest, Monty and Reggie, were returning to The King's School at Parramatta after the summer holidays. With them were George and Percy, who were taking them to Goulburn to catch the coach.

They had just turned into the main road at the end of the drive when they were hailed by Ben Hall, the bushranger. He and two companions, John Gilbert and Johnnie Dunne, had already robbed two mail coaches that morning and now thought that the four-in-hand and its horses would complete a profitable day's work.

The boys had other ideas. George whipped the horses to a gallop and, as the bushrangers opened fire, he turned off the road to cross the paddocks back to the homestead as his

Jim Maple-Brown, great grandson of William Pitt Faithfull and the present owner of Springfield.

Pamela Maple-Brown.

brothers returned the fire. Gilbert took aim at full gallop but as he fired, his horse raised its head and was killed instantly.

When the four-in-hand reached a fence Percy ordered the boys to run for the homestead whilst he covered their retreat with his one remaining shot. He aimed it at Gilbert, who had been thrown as his horse fell and who was now running along the fence. The shot hit one of the posts and Percy took off in a hurry as Ben Hall spurred his horse in pursuit. All the boys reached the homestead safely and the bushrangers withdrew.

Some ninety-four shots had been fired in only a few minutes and the only casualty had been Gilbert's horse. Impressed with the boys' courage, the New South Wales government presented them with a gold medal which is still kept in the homestead at Springfield.

By the time William Pitt Faithfull died in 1896 another of his sons was ready to take over. Lucian Faithfull had been born in 1855 and after leaving Sydney Grammar School he went to a property at Mudgee owned by Richard Cox, who was acknowledged as one of the best sheep men of his day. When Lucian returned to Springfield in 1871 to take over the management of the stud flock he had two clear objectives. The first was to breed a pure stud flock of the finest merinos. The other was to breed a separate flock of stronger, heavier-woolled sheep by carefully selecting them from the stud flock.

Refusing to have anything to do with the wrinkled Vermont sheep, Lucian developed a highly regarded stud that soon rivalled any in the country. When he died in 1942

he left the property to his daughter Bobbie, who in 1923 had married a local grazier called Irwin Maple-Brown.

They moved into Springfield and after the war started the long job of restoring the property. Fences were built, tanks made, buildings renovated and, equally important, the Springfield stud was slowly built up again.

In the meantime they had acquired another property not far away called Fonthill and their son, Jim Maple-Brown, was sent to run it. He was already developing his own views about sheep breeding and it was thought that he could try them out at Fonthill without damaging the famous Springfield stud. In 1964, when his father died, Jim Maple-Brown returned to Springfield, but the two studs were still kept separate.

Jim Maple-Brown was not a traditional breeder of sheep, even though he came from one of the oldest families in the business. He thought that the merino that was being bred in Australia was no longer suitable for the market conditions that were emerging. It was a magnificent animal but it had one drawback for the commercial grower: it had been developed with the main emphasis on wool production. When meat became more profitable than wool the merino breeder was unable to convert to meat production without introducing British breeds, which increased the fibre diameter of the wool and reduced its uniformity.

The sheep of the future, he thought, should be a merino that was capable of producing meat as well as first-class wool and thus allow the grower to change emphasis as the market changed.

In 1965 he read of a South African strain called the Letelle which seemed to offer some of these characteristics and two years later he went to have a look at it. The trip did much to crystallise his ideas. He realised that it would be possible to breed a multi-purpose sheep and still retain the traditional fineness and uniformity of Australian merino wool. However, high labour costs meant that it would have to be plain bodied and easy to manage if it were to survive economically.

When he returned he started to select sheep with these characteristics from the Fonthill stud. Ten years later he had succeeded in producing sheep of exactly this type and the nature of the whole flock had changed. Thinking that these sheep would produce less wool he had continued to keep the two studs separate, but when repeated tests showed that the wool weight was no different he took the major step of integrating them. Knowing that the name of Springfield would always be associated with the traditional type of merino, he called the new stud Fonthill.

During this lengthy period of development and breeding Jim Maple-Brown realised that a computer could be used to increase the efficiency of selecting for those qualities that he wished to develop. In this he was helped by Dr Helen Newton Turner, a widely acclaimed geneticist who had already shown that some traditional methods of evaluating merinos were not always reliable.

Crimp frequency in wool, for example, had long been used as an indication of its fineness, but Dr Turner maintained that it was not only an unreliable guide, but that it was positively misleading. She pointed out that heavy cutting sheep tended to have fewer crimps in a given length of fibre than lighter cutting ones even when the average diameter was identical.

By now Jim Maple-Brown was convinced that the breeding worth of sheep could be assessed on their relative measured production. In addition, the computer could store, process and retrieve the considerable amount of data involved and apply it to produce a more efficient system for selecting sheep. With further help from Dr Turner he started to refine the techniques.

There were two fundamentals. One was that all the important factors of a sheep could be measured with considerable accuracy. The other was that those factors were also highly heritable, so that superior animals would pass on much of their superiority to their offspring. The measurements therefore provided a clear assessment of a ram's relative worth not only as a producer, but also as a source of genes for improving the flock.

For example, the measured fineness of wool is much more relevant to sheep breeding now that it has become the accepted method of assessing wool at auctions. The value of a sheep's wool production depends largely on the weight of clean wool produced and its average fibre diameter. Weight is the dominant factor but fineness is also important because historically it has always commanded a premium. In the five years ending June 1982, the average price of 20 micron wool

Classing rams in the old woolshed at Springfield. Jim Maple-Brown, top, operates the computer whilst his son, Richard, below, checks the screen for details of the ram he is examining.

The elegant homestead at Springfield. Dating from the 1850s, it has nearly forty rooms — although it is a while since anybody counted them.

Some of the permanent water provided by the Mulwarree Chain of Ponds.

The sitting room in the Springfield homestead.

was 4 per cent more than that of 21 micron. Indeed, over the whole 19 to 24 micron range, a difference of 1 micron changed the price by a little over 4 per cent.

The value of a sheep's wool production can therefore be increased by selecting either for increased weight or smaller average diameter, or a combination of the two. The computer makes it relatively easy to incorporate these factors into a selection system that will more accurately assess the breeding worth of each ram.

At Springfield the basic data used to select each ram is the weight of clean wool it produces, the average fibre diameter, its body weight and its identifying number. The first three are combined in a formula developed with the assistance of Dr Turner which is designed to increase by equal percentages the income generated by the sales of meat and wool.

The selection process for rams starts in September each year when they are about twelve months old. Those with obvious faults such as excessive wool on the face or structural defects are culled from the mob. The rest have numbered metal tags placed in their ears for permanent identification and they are then shorn and the fleeces weighed. A sample is taken from the mid-rib area of each fleece and sent to the University of New South Wales, where it is used to calculate the average fibre diameter and the yield of clean wool for the whole of the fleece. In late December, when this information has been received, the rams are again checked for visual faults and reweighed.

The complete data for each ram is now entered into the computer at Springfield and, using a formula which accurately reflects the breeding objectives of the whole flock, it calculates a score for each ram. It also calculates the average production figures for all the rams born that year and this information is used to monitor the efficiency of the selection system and the rate of genetic progress being achieved.

The computer then lists the rams in order of merit and this list is used as a basis of deciding the various grades. For example, the sires to be used for stud use might be all those rams with a production index greater than 118 per cent of average, whilst the cut off points for the sale grades and the cull level can be located at appropriate points further down the scale. The computer can then show the grade of each ram, together with all its other information, against the number on its ear tag.

It is 8.30 in the morning and in the woolshed they are about to class some 900 rams. About 600 of them are already in pens inside the shed, whilst the rest are held outside until there is room for them to come in. In the middle of the shed is a race wide enough to take only one animal. Whistling and waving their arms, the men drive some sheep from the nearest pen into a smaller pen leading to the race. Others move them into the race itself, the feet of the sheep clip-toeing on the slatted floor as they form up like refined old ladies at a department store sale.

The old coach house.

At the head of the race a man reads the ear tag of the first ram and calls its number to Jim Maple-Brown, who is in a small cubicle behind him. He punches the number into the computer and full details of the ram come up on two screens, one inside the cubicle for the computer operator and the other immediately above the race. Richard Maple-Brown, Jim's son and now manager of Springfield, looks at the details on the screen and then looks at the animal. He can see no reason to alter the computer's grading and he calls a confirmation back to the cubicle. As this is being entered into the computer, Richard marks the ram with a coloured dye to indicate its grade and releases it from the race into the appropriate pen. The sheep in the race shuffle forward and the next number is called out.

This time Richard thinks the animal has faults beyond those attributes measured and displayed by the computer, so he down-grades it and calls out the information. The essential rule is that an animal can only be down-graded — it can never be moved up a grade no matter how good it might look. As the hours go by, sheep are moved round the pens inside the shed and others are brought in from outside. The pens holding the culls start to fill, for this will be the biggest category of all.

By the end of the day they have classed all 900 rams. The previous day they had used the computer print out to select 80 top rams for stud use and that pen should now contain those 80 sheep. It does. They will all have the characteristics needed to advance the stud's genetic pool in the right direction, and they will have more of those required characteristics than the other 820.

When a buyer comes to select rams he too looks at the details on the screen as he examines each animal. He can see the diameter of the fibre, the wool factor and the body weight and how they compare with the average of the generation of rams he is inspecting. Provided he knows what he wants, he can select his rams without guesswork. Reluctant at first, the buyers now rely on this information more than their own assessment. In the process they buy more than 400 rams each year.

If modern technology has been brought to the old brick woolshed that has been standing for more than a hundred years, the contrast is no less in the homestead. A fine, elegant building dating mostly from the 1850s, it is a magnificent house. Antique furniture is everywhere but the big windows keep the rooms light and pleasant. There are between thirty and forty rooms, but nobody is quite sure. It is a long time since they counted.

In the oldest part of the building a suite of rooms has been set aside as a museum. One of them is furnished in the Victorian style, or rather Victorian furniture has simply been left there. On the tables are the knick-knacks of Victorian family clutter, relics of the bushranging episode and assorted *objets d'art*. It is not a museum, it is alive. Lucian Faithfull could walk in at any moment and wonder what I was doing.

Fonthill merinos.

But at the other end of the house, in his office just inside the front door, Jim Maple-Brown is watching the computer print out a record of the day's work. The office is a cluttered place with old show ribbons lining the wall and even hanging from pictures. Files are piled up against the book shelves and stacked on old, comfortable chairs. The computer terminal throws out a blueish light as it talks to itself in an endless clatter and the paper neatly folds itself into a waiting basket.

That evening Bobbie Maple-Brown comes to the homestead to celebrate her eighty-third birthday. She lives in a modern house about half a kilometre from the homestead, stoutly maintaining that she has not retired.

The family is assembled in the drawing room as a car glides up the gravel drive and sighs to a halt outside the big french windows. Bobbie comes in cursing the stick she has to use to climb the steps and settles herself into a settee that she has

known all her life. She is trying to give up smoking and says that she has brought with her only three cigarettes and three matches. When she unwraps the present from Mary, the cook, she is delighted to find two packets of cigarettes but is left with the problem of having only three matches. I give her mine.

In the room are four generations of Maple-Brown Faithfulls. From the time the property was founded in 1827 only the first two generations are not present: William Pitt Faithfull and his son Lucian.

In the dining room the huge mahogony table is lit by candles and the light barely reaches the corners of the room. The jackeroo tries to work out which knife and fork he should use for each dish and Bobbie tells a story that would take any radio station immediately off the air.

And through the door comes the sound of the computer, still talking to itself as it settles down for the night shift.

HADDON RIG

A working day comes to an end at Haddon Rig and men are returning to the neat white buildings near the homestead. Near the big woolshed beyond the tree-fringed lagoon a long plume of dust stretches behind a ute as it moves quickly across the paddock. A few minutes later it stops in front of the small office and a lean man jumps down and goes inside. He looks almost young enough to be a jackeroo, but not quite. Fit and well tanned, he might be a supervisor returning from a long day in the paddocks. But he is neither jackeroo nor supervisor. He is George Falkiner and at twenty-seven he is the owner of Haddon Rig, one of the outstanding merino studs in Australia.

Haddon Rig is about 160 kilometres west of Dubbo in New South Wales and 32 kilometres from the small town of Warren. There are many sheep studs in the district, but there are other kinds of farming as well. South of Warren huge wheat fields stretch away on either side of the road, whilst close by are lush green paddocks growing cotton. But once beyond Warren, on the road to Quambone, the country takes on the unmistakable look of sheep country. It is flat and lightly timbered and grows a good cover of tussocky Mitchell grass on which sheep thrive.

Haddon Rig consists of 64,500 acres of red and black soil country. It is a good combination because the black soil grows winter feed that is particularly rich in protein. The Marthaguy Creek runs through the property and provides an almost constant supply of water which is supplemented by bores, ground tanks and irrigation. The rainfall is about 18 inches a year and it can fall at any time, although recently it has not been falling very often. But a few days ago they had 160 points and although it did not break the drought it was enough to put a green haze on paddocks that had previously been brown and arid.

They run about 32,000 sheep at Haddon Rig and each year cut 900 bales of medium wool of about 22 microns with a characteristic brilliance and soft handle. Each year they sell about 3,000 rams to clients in most parts of the country, and it is the quality of these rams that carries the reputation of this famous stud.

Some fetch enormous prices at auction and the news is carried in the farming press and sometimes even in the city dailies. But outstanding though these sales might be, the majority of rams at Haddon Rig are sold to commercial growers who arrive once a year to select a handful of flock

George Falkiner, the twenty-seven-year-old owner of Haddon Rig.

rams at prices ranging from just over $100 to nearly $500. It might be less newsworthy but, like most sheep studs, it represents the bulk of the business.

Haddon Rig has not always been what it is today for when this land was first taken up the Falkiner family had not yet started the dynasty that was to have such a profound influence on Australian sheep breeding. And when Franc Sadleir Falkiner did buy his first sheep station it was Boonoke he chose, not Haddon Rig. Nor, indeed, has Haddon Rig always been a sheep station, for in the early days it was predominantly a cattle run.

The first owner of Haddon Rig was William Charles Wentworth, who in 1813 had discovered the way through the Blue Mountains with Lawson and Blaxland. It was either Wentworth himself or his partner Christie who called the property Haddon Rig after a hunting lodge owned by the Duke of Buccleuch in Scotland. By 1868 Wentworth had about fifteen properties in the Warren district and that year he applied to the government for the issue of Crown leases.

N

Marthaguy Creek

Irrigated area

Braémar

Irrigated area

Irrigation Channel

Irrigated area

To Quambone

HOMESTEAD

To Warren

Weemabung Creek

Airstrip

Two years later he sold Haddon Rig to his son Fitzwilliam. But the following year the Macquarie River rose to a record level, inundating the country and drowning most of the stock.

Unable to carry on, in February 1873 Fitzwilliam Wentworth sold Haddon Rig to a Scot called James Richmond. A very experienced grazier, Richmond had already built up a number of properties in Victoria and had established a reputation as a breeder of fine sheep. Later that year he overlanded 8,000 ewes from his property at Corowa on the Murray River and then started the daunting job of fencing Haddon Rig. Two years later he had erected about 345 kilometres of boundary and paddock fences and in the process had used almost the entire stock of fencing materials held in Sydney and Melbourne.

Richmond worked energetically to improve Haddon Rig and spent many thousands of pounds on it. He built an enormous woolshed of pine and shingle that could hold 1,500 sheep and which soon became one of the show pieces of New South Wales. But it was also functional — in 1875 Haddon Rig shore more than 74,000 sheep and produced 600 bales of wool.

It was a remarkable achievement for in only two years he had taken a run-down cattle property and had turned it into one of the leading wool producers in New South Wales. True, he had spent a considerable amount of money in the process, but he now had a profitable station that was worth considerably more than when either of the Wentworths owned it. His profits too were carefully used and the following year he started to buy the freehold for £1 per acre.

Although the property was now a major wool producer, it was not yet a stud. It was not until 1882 that Richmond started the Haddon Rig stud with 30 rams and 1,900 ewes which he bought from Austin and Millear at Wanganella. Most of the rams were sired by two outstanding Wanganella rams, Premier and Warrior, and from them Haddon Rig quickly established its reputation for large sheep with heavy fleeces of medium wool. He bought more stock from Wanganella in the following years but since 1894 no imported rams have been used at the Haddon Rig stud.

Like many graziers at that time, Richmond strongly resisted attempts by shearers to introduce unionism into Australian woolsheds. He refused to shear on union terms and in 1890, as a result of threats made by some of the shearers, he had 30 policemen guarding the homestead. They were in the wrong place, however, for on the night of 16 October Richmond's large and famous woolshed was burnt to the ground and 1,500 sheep, penned for the following day's shearing, died in the fire.

Richmond immediately brought non-union shearers from New Zealand and continued the work in a neighbour's shed. A new shed, still standing at Haddon Rig, was built in 1891 and three years later they shore 133,000 sheep. The drought which had started two years earlier was at last coming to an end and Richmond was able to sell 12,000 wethers for 7/6 each, the best price they had seen for some time.

Charles Nicholl, who managed Haddon Rig for many years before it was bought by Bert Falkiner in 1916.

But by 1902 the country was facing one of the worst droughts on record and Haddon Rig fought a constant, and expensive, battle to keep sheep alive. By April they were hand feeding 28,000 sheep and this was costing £3,000 every month.

Until then, rabbits had been increasing at an alarming rate. The drought reduced their numbers considerably, but when the good seasons returned, so did the rabbits. Some paddocks had already been fenced with wire netting before the drought and the rabbits had been eradicated from these areas, but after the drought they killed no less than 370,000 rabbits in one year. It was an endless task.

By now the property was being managed for Richmond by his nephew, Charles Nicholl, but his reports were becoming increasingly gloomy. In 1916 Nicholl was once again describing the effects of drought. The Marthaguy Creek was running dry and they would soon have to start carting drinking water to the homestead.

Richmond, who was now eighty-two years old and living in his native Scotland, decided it was time to sell. An auction was announced but the drought did little to encourage buyers and only one person arrived at Haddon Rig to inspect the property. It was Bert Falkiner.

The woolshed at Haddon Rig, built in 1891 after the original shed had been destroyed by fire.

The eldest son of F. S. Falkiner of Boonoke, he had run that property for a year or two after the death of his father before leaving to look for a property of his own. Now he had found it. Although not looking its best, Bert Falkiner bought Haddon Rig before the auction for £135,000.

For Charles Nicholl it was the end of an era. He wrote to his uncle in Scotland:

I have had a long innings with you, nearly thirty-four years it will be on 1st February. I have seen many changes and wonder what the next one will be. I hope the old stud goes along as well as in the past, if so few will equal or come up to her...

There were indeed changes, and not all of them to Nicholl's liking. With business interests keeping Bert Falkiner in Sydney, Nicholl stayed on as manager, but the changes were immediate. The homestead was improved and new account-ing methods introduced to replace the less-than-perfect systems of the past. Nicholl was told to cull out all the wrinkly sheep to prevent fly strike and to reopen hostilities on the rabbits, which had made a new appearance.

To Nicholl, every change carried an implied criticism and, unhappy at what he saw as interference, he soon resigned. Bert replaced him with George Griffin, then manager of another Falkiner property near Boonoke, but he also realised that if Haddon Rig was to retain its reputation he would have to spend more time there himself.

He applied the same philosophy that had been so success-ful at Boonoke. Even when seasons were good he took great care to avoid overstocking the property. As the flock grew he bought adjoining land to accommodate it. By the middle of the 1920s he had bought three more properties and his holding had grown to about 84,000 acres. One of these properties, Boomanulla, consisted of 4,000 acres along the banks of the Macquarie River.

When Bert died in 1929 Haddon Rig was taken over by his elder son George, then a twenty-two-year-old engineer-ing student at Sydney University. Young and energetic, George soon started on an ambitious plan of improvement. Buildings were renovated, fences were replaced and the rabbits finally conquered. In 1936 he pulled down the old homestead and built the present one on the same site. By the start of the Second World War he had already invested more than £45,000, and the work started again as soon as the war was over and men and supplies became available. He built a huge tank to hold 36 million litres of irrigation water and installed a generating plant to supply electricity to the entire homestead area. As a result, his station hands were enjoying the benefits of electric refrigeration long before most people in Sydney or Melbourne.

He had problems, too. In 1955 more than 11 inches of rain fell in one night. The Macquarie River rose at an alarming rate and by the morning it had joined up with the

The homestead, built in 1936.

Marthaguy Creek several kilometres away. Although air drops and good organisation saved most of the stock, it made him aware of the importance of good water management as a protection against both flood and drought. It is a lesson that has been remembered ever since.

When George B. S. Falkiner died in 1961 at the age of fifty-four many sheep men in Australia and overseas wondered what would become of Haddon Rig. His son, also George, was only six years old and there seemed every reason to believe that this famous stud might now be sold. There would have been no shortage of buyers this time.

But it soon became known that the property was not for sale and that it would continue to trade as usual. Fortunately for the family, the company board was at that time extremely competent. It was headed by John Brown, a Scot who had served with distinction during the war before coming to Australia in 1948. He had joined the board ten years later and it was under his guidance that the company was able to continue until George Falkiner was old enough to take over.

It was not an easy time. The droughts of the Sixties were followed in the early Seventies by a run of very low wool prices. In 1972 there were over 2,000 unsold rams on Haddon Rig and in the end most were sold to the meatworks for a dollar a head.

Meanwhile, George Falkiner went to school at The King's School in Parramatta and then took a degree in accountancy at the University of New South Wales. He realised even then that his future role would have more to do with financial management rather than the day-to-day running of the property.

He joined the company in 1976 and made an extensive trip to north and South America to investigate irrigation techniques. In 1980 he spend six months on cattle properties in the Northern Territory. 'I was there to gain experience, of course,' he says, 'but there was another reason. I wanted to see if I was good enough to survive.' He was.

When he took over the running of the company in 1981 his financial training was immediately useful. The transition took two years and during that time he introduced computerised office systems and reorganised the company structure.

The company now owns four properties. There is Haddon Rig itself, centre of the famous stud built entirely on Wanganella Peppins. Then there is Strathaven, near Gnowangerup in Western Australia. This property of some 15,000 acres was bought in 1963 to develop the stud business in that State by producing sheep more suited to its climate. Now managed by Ken Littlejohn, it has been extremely successful. In 1982 Strathaven paid $27,000 at the Adelaide Ram Sales for a Collinsville ram called JC & S 80. It has remained on Strathaven ever since but frozen semen has been sent back to Haddon Rig to start a new family of bigger sheep that will retain the traditional Haddon Rig wool.

Forbes Murdoch, manager of the Haddon Rig properties.

Andy McLeod, assistant manager at Haddon Rig.

The third property, only a few kilometres from Haddon Rig, is Canonbar at Nyngan. It was bought in 1982 and consists of more than 23,000 acres running 12,000 stud sheep. The flock was founded in 1859 and rams from Wanganella and Boonoke were used until 1935. Since then it has used nothing but Haddon Rig rams.

Finally there is the farming and irrigation activity that occupies part of Haddon Rig itself and nearby Boomanulla. There, a massive irrigation line 800 metres wide moves itself slowly across a long paddock, drawing water from a channel leading from the Macquarie River. There they grow lucerne, wheat and soya beans — some as cash crops, others as feed for the stud.

The Haddon Rig properties are managed by Forbes Murdoch. Born in 1942 on the Queensland sheep station of Terrick Terrick, he was educated at The Southport School. In 1959 he started as a jackeroo at Athol, near Blackall, and in 1960 moved to Haddon Rig, still as a jackeroo. In 1963 he was sent to Western Australia to manage the recently acquired property of Strathaven and his ground work there did much to ensure its later success. In 1975 he became manager of Haddon Rig and joined the board the following year.

High revving and seemingly tireless, Forbes Murdoch lives in a state of perpetual motion and spends most of his time well away from the tiny office near the homestead. Managing a property of this size is an active life and the days when such men wore jackets and ties are long gone.

Livestock photographer Les Jones at Boomanulla.

Haddon Rig rams.

Helping him as assistant manager is Andy McLeod. Born at Coleraine in Victoria, he jackerooed for two years at Nareen, owned by Malcolm Fraser, before moving to Haddon Rig thirteen years ago. An ex-Collingwood footballer, Andy McLeod still has the fitness of an athlete and puts it to good use as he organises the ten jackeroos. Although most of them are twelve years younger than he is, none of them doubt that he can more than match them in the hard physical work of running the property.

With the day-to-day running of Haddon Rig in good hands, George Falkiner spends about half his time in his Sydney office. There he is involved in financial planning and the overall control of all the properties. It is a vital job, for this is a big business and it has to be run like one. Cash flows, trading accounts and computer projections are essential and they form the basis of many decisions which in turn might have a profound influence on the whole of the Australian sheep industry.

But no property owner can spend all his time behind a desk, or want to, and George Falkiner is no exception. Using his blue and white Cessna 182, Mike Lima Yankee, he leaves Sydney regularly to visit clients in New South Wales and Queensland before returning to the real base at Haddon Rig. For it is there, on the black and red soil of Warren, that they breed the sheep that has made Haddon Rig famous throughout the world.

Now, at the end of the day, he walks into the office and there is an almost instant meeting with those already there. Dick Jago is in from Dubbo to buy rams for one of his clients. One of the best sheep classers in the country, he classes the Haddon Rig stud by the traditional method of inspecting each ram and making a decision about its quality and potential. It is a heavy responsibility, as the progeny of these rams will be used in flocks throughout most of Australia.

Forbes Murdoch joins in the discussion and does three other things at the same time, and Andy McLeod keeps an eye on the jackeroos returning from work as he draws up plans for the following day. Don, the bookkeeper, takes a phone call from a distant client and the base radio suddenly squawks as a man checks in from a far away paddock.

It is 7.30 the following morning and at the ram shed at Boomanulla Les Jones is getting ready to photograph the rams. The open-sided shed is built of hand-hewn posts and logs and stands in one corner of a very green and lush paddock on the banks of the Macquarie River. In the middle of the paddock, waiting curiously for the action to start, are the rams. They are the state of the art of Haddon Rig merino breeding. Large, handsome and aloof, they have been brought to perfection for the Dubbo Ram Sales in three weeks time. Until then they will live like lords, looked after by Mick who lives alone in a small white cottage at the far end of the paddock and who looks after room service.

Les Jones is an ex-journalist who has since turned to the demanding job of livestock photography. He lives on a 40-acre property near Dubbo and works without a studio. He

has no need for one, for all his subjects are in paddocks like this. He comes to Haddon Rig once a year to take advertising pictures before the ram sales and they have to be good. Many of these rams will sell for $16,000 or more and nobody will be happy if the photographs make them look ordinary.

The rams have grouped themselves near a big gum tree in the middle of the paddock. This and other trees nearby are throwing long shadows across the rams and Les asks Forbes Murdoch if he would move them into a sunny part of the paddock. He goes over and prepares his cameras as Forbes and Mick gather the sheep. With absolute confidence they reassemble them and then the rams wheel and turn with Guards-like precision until they are walking line abreast towards the camera.

Les Jones works his Nikons like a man who could do it in his sleep. The sheep stop, are taken back, reformed into a group again, and now the group turns as the first of the rams form the line. Other progressively extend it until once again they are all walking towards him line abreast. It is so quiet you can hear the shutter tripping. Then they do it all over again.

Satisfied at last, they start the more difficult job of photographing the six best rams. Each one has to be photographed by itself and each must be carefully positioned so its quality is brought out. Mick and Forbes stand the first ram in front of the green of a willow tree. Forbes bends down and moves the hind legs so that one does not cover the other. Les lays flat on the ground and looks through the camera as Mick does the same with the front legs.

'Far hind leg forward a little,' he calls. 'That's enough.' Forbes backs away gently, leaving Mick holding the ram's chin. 'Ease his head back, Mick. Now turn it a bit towards me. Fine. Now lower his chin — that's it. Fine.'

The ram moves his feet. Forbes walks up and they start all over again.

It takes them two hours to photograph the six rams. Some seem to know what is expected of them and stand proudly in the correct position. Others are not so easy and need a lot of help and encouragement. For many people it would be a time for frayed tempers, but these men know that would only make matters worse. Les works quietly and patiently, waiting for the animal to look just right before he takes the picture. And Mick and Forbes have to rely on his judgement because he is the only one seeing the animal as the camera will see it.

When they have finished Les goes back to the homestead to improvise a darkroom. By the time he leaves a few hours later he has developed the negatives and printed what he thinks are the best so that the men of Haddon Rig can study them at the end of the day. They have to be right, like everything else at Haddon Rig. There is no room for second best.

Meanwhile, George Falkiner is several kilometres away on part of the property called Braemar, selling rams to a buyer who has arrived on his annual visit. The rams are passed through the race and the buyer examines each one carefully, parting the wool on the side of each animal to judge its

Part of the screened verandah which extends around all four sides of the homestead.

quality. The rams he selects are collected in a pen and then he goes through them again to make his final choice.

Other buyers arrive throughout most of the day and George Falkiner looks after all of them whilst the jackeroos move the rams into the pens. It is hot and tiring work but everybody knows how important it is.

At the end of the day George Falkiner drives back to the homestead and pauses at the door to take off his boots. There is nobody there except the housekeeper and the big homestead, 1930s baronial, is quiet and empty. His mother died in 1977 and his two sisters now live in Sydney, so that it is not really the family home any more.

He collects a beer from the bar and takes it on to the long verandah that runs down the side of the house. The sun lights up the windmill nearby and turns it into a golden sculpture whilst the ducks call to each other along the banks of the creek in natural stereo.

For George Falkiner, the twenty-seven-year-old owner of Haddon Rig, it is the best part of the day. Until the phone rings.

THE ROYAL FLYING DOCTOR SERVICE

A hundred kilometres from the homestead at Nappa Merrie in south-west Queensland, near the ruined outstation of Baryulah, is a small and lonely graveyard. The headstones, many broken and partly covered by sand, are difficult to read now for time has scrubbed them to a smooth cleanliness. But one piece is still legible. Faint words carefully carved in the soft stone record the death of a nineteen-year-old stockman who, many years ago, was thrown from his horse. Far away from help of any kind, his death would have been almost inevitable.

There are many graves like this on outback properties. Some, close to homesteads, are still well tended; others are on parts of properties that are still remote and seldom visited. Together they bear testimony to the dangers the people of the outback faced, dangers that they accepted as an inescapable consequence of being there.

As recently as 1912 there were only two doctors serving an area of 1.8 million square kilometres in the Northern Territory and Western Australia, and other parts of the outback were little better. Church missions did what they could to provide additional medical help, but in truth they could not do very much. Travelling was slow and difficult for missionaries and patients alike — and at times it was simply impossible. Distances were great, roads were often impassable, and people died.

The situation today is dramatically different. The Royal Flying Doctor Service, through a network of fourteen bases, takes medical services into all parts of the outback with a speed and efficiency that even now seems little short of miraculous.

It was started through the efforts of John Flynn, who in one dedicated lifetime did more for Australia than many of those who have more famously dominated its history. A modest and unassuming man, his name is forgotten by most Australians now. But it is still remembered in the outback, for he brought hope and safety where none had existed before.

John Flynn was a missionary with the Australian Inland Mission (AIM), an organisation run by the Presbyterian Church to take religion and comfort to isolated areas. He was the first to see the value of aviation and radio in the outback and campaigned relentlessly to establish a medical service that would combine the benefits offered by these new, infant technologies.

Eventually he was successful and in 1928 the AIM Aerial Medical Service was started at Cloncurry in Queensland. It used a pilot and aircraft provided by a small bush airline, Queensland and Northern Territory Aerial Service — later known as Qantas — and provided radios to isolated communities which could be used to call for medical help. Designed by a brilliant young technician called Alfred Treager, the radio had a keyboard that transmitted the correct Morse signal when each letter was pressed and ran on power from a generator fitted with bicycle pedals which was worked by the operator.

The Service was so successful that the AIM eventually handed control over to a national body set up for that purpose, and this became the Royal Flying Doctor Service that we know today.

Its importance to outback communities can hardly be overestimated. Sheep and cattle stations, mining camps, Aboriginal settlements, all can be so far from established medical services as to make their use little more than theoretical. They are too far away for minor treatments, and for emergencies too. But the need for these services is as great in these communities as it is elsewhere, and it is the Royal Flying Doctor Service which supplies them.

They understand the situation well. John Hepworth, director of the base at Alice Springs, says, 'Isolation is never greater than when people are sick — and in spite of all our technology the need for help is as great now as it ever was. Technology might change, but people don't. Being sick in an isolated settlement can be a frightening experience.'

There is no doubt about that, for the climate and isolation can add their own complications to make a relatively minor illness very serious indeed. Diarrhoea in a child, for example, can lead to a rapid dehydration that is more dangerous than the sickness that caused it. It needs expert attention, and very quickly, if the child is to survive.

Accidents too can present major problems. Fractures or serious lacerations are painful enough in the city, but at least help is usually only minutes away. In a distant

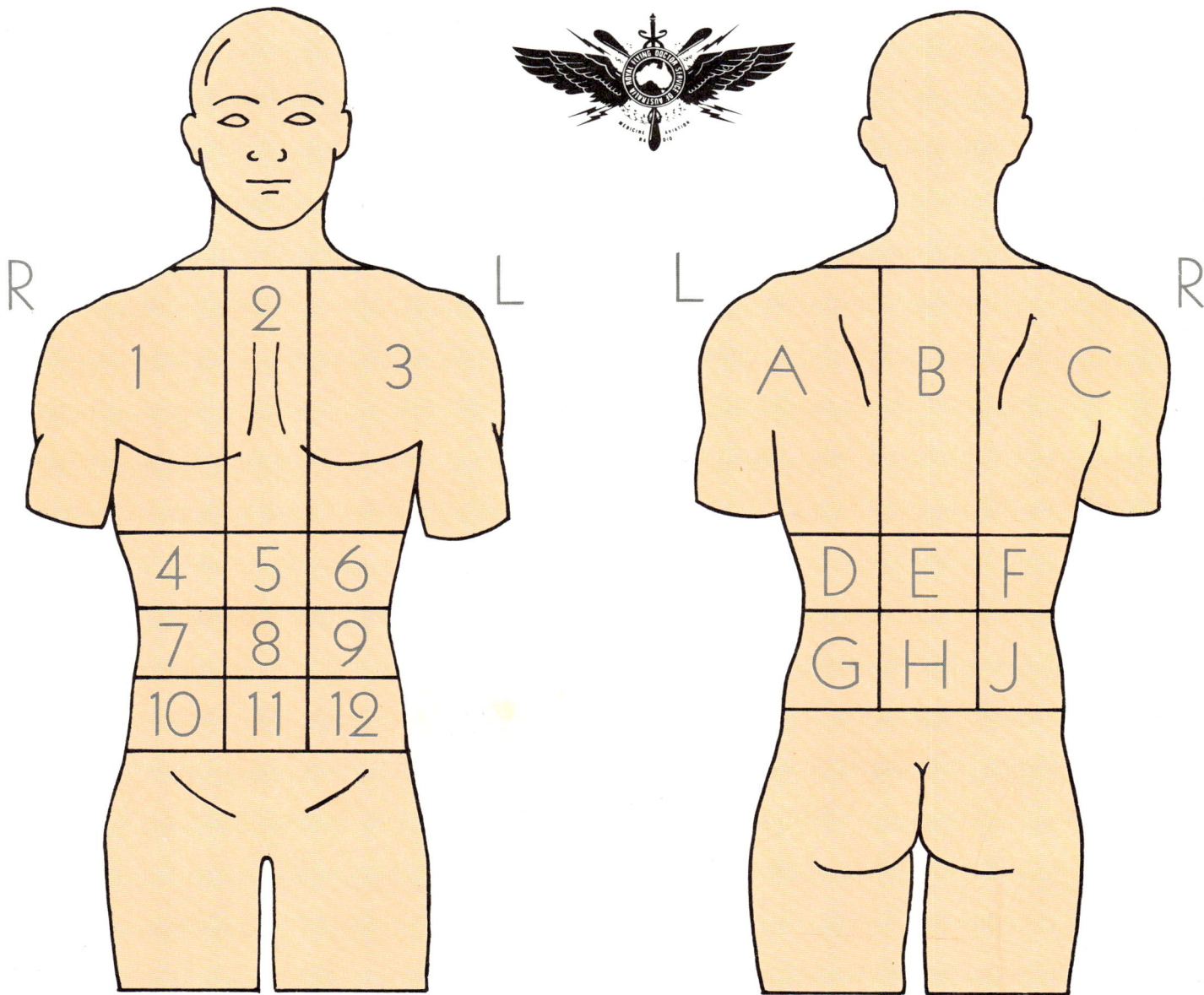

Diagrams of the human body used by the doctor and patient during a radio consultation.

part of a property, perhaps a hundred kilometres from the homestead, it might take more than two hours to get help and even then it is unlikely to be much more than basic first-aid. If a man has several fractures and is bleeding freely he needs much more than that.

These are still potentially dangerous situations, but the Royal Flying Doctor Service reduces these dangers and, almost as important, takes away much of the fear associated with them. The distressed mother and the injured worker now know that skilled help is on the way and that instructions given over the radio will make sure the correct action is taken until it arrives. There is still tragedy, to be sure, but it is no longer born of helplessness.

The base at Alice Springs, established in 1939 to cover an area of some 1¼ million square kilometres of the Northern Territory, now serves a remote population of some 30,000 people. It operates three Piper Navajo aircraft, each with full medical equipment, and is staffed by four doctors, three pilots and five communications experts.

Whilst the dramas might make exciting reading in the city dailies, it is the constant care that the Service maintains, all day and every day, that is more important. Every day the base at Alice Springs conducts a routine medical consultation over the radio from 10 o'clock in the morning and people on distant communities use the radio to consult the doctor. The doctor cannot see the patient, but the patient describes the symptoms and refers to a diagram of the human body which is divided into numbered sections, and so can tell the doctor fairly accurately where the problem is located.

Having decided on the treatment, the doctor will prescribe drugs or medication which are kept in a specially packed medical chest at every distant settlement. These are also referred to by number to avoid the confusion that could arise from the use of unfamiliar names.

If it is routine to the medical staff at the base, then so is much of the flying. Each month doctors and other specialists fly to most of the distant settlements to hold clinics. There they can examine patients whom they might have treated over the radio and make routine checks on recently born babies and their mothers. People will often travel hundreds of kilometres to attend the clinic with a problem that might have seemed 'too minor' to justify a radio consultation, or to have a check-up for a continuing condition such as diabetes or arthritis.

But important though this work is, it is the sudden emergency that draws most heavily on the very considerable skills of the Service.

An emergency call can be made to the base at any time of day or night. If the base is not manned, the incoming call will be automatically switched through to a doctor on call, wherever he might be. The result is that the caller will be talking directly with a doctor within a matter of minutes. 'Almost certainly a lot quicker than you would in a city during a weekend, or probably any other time,' says John Hepworth.

If the doctor decides that the patient needs immediate treatment, and perhaps evacuation to hospital, he will call the duty pilot for an immediate flight. He will be airborne within twenty minutes, but meanwhile he will give instructions for emergency treatment over the radio and will continue this from the aircraft.

If weather conditions make flying dangerous, and if he is convinced that it is a matter of life or death, he will ask for a 'mercy flight'. A mercy flight is one which might result in a reduction of the normal standards of air safety, and as such the decision whether to make it or not rests entirely with the pilot.

The risks can be considerable. The flight might be through stormy conditions at night and at the end of it the pilot might have to land on a seldom-used airstrip that has no navigation aids, is probably dangerously short, and which is lit by a few car headlights. Having landed, the doctor will have to carry out emergency treatment in the middle of nowhere and then continue it during the no-less dangerous flight to the nearest hospital, perhaps hundreds of kilometres away.

It is amazing that people do it at all, but they do — and fairly often. It is a reminder that the outback can still be a dangerous place and that its isolation can never be taken for granted. Flynn, with his vision, brought hope to the outback, and the Royal Flying Doctor Service, which now treats nearly 100,000 patients every year, is now fulfilling it.

The pity is that it came too late for that young stockman at Baryulah.

Base Director John Hepworth in the radio room at the Royal Flying Doctor base at Alice Springs.

NAREEB NAREEB

If Hamilton, in the Western District of Victoria, exaggerates a little when it claims to be the wool centre of the world, it is only a little. For the Western District is one of the outstanding wool-producing areas of Australia and has been for generations.

About 60 kilometres east of Hamilton, along a country road lined with paddocks and an occasional lake, stands Nareeb Nareeb. It consists of about 6,700 acres of gently rolling country, some of it rich loamy soil and the rest basalt. To the north the Grampian mountains rear up from the plain and dominate the skyline, hazy blue in the distance.

Nareeb Nareeb looks totally at ease with itself and its surroundings, as if nothing could disturb the peace of such a well-bred and affluent district. It is a kind of rural Toorak which somehow guards itself from the more unpleasant realities of life. But the tranquillity is deceptive, for Nareeb Nareeb has faced more harsh realities than most and its gentleness has been shattered on more than one occasion.

Charles Gray was born in Anstruther in Scotland in 1818. His father was in the Royal Marines and it was assumed that Charles would follow a similar career. But when the time came he preferred the uncertainties of life in Australia to waiting for promotion in Britain that was then relatively peaceful.

He arrived at Port Phillip in 1839 and whilst spending a few months on the property of a friend on the River Leigh he heard of a small flock of sheep that was soon to be offered for sale. He wrote to two men he had met on the voyage out, Scott and Marr, and together they bought the sheep and squatted on good land on either side of a creek. It was the start of Nareeb.

The only tools they had were a broken axe and an old trowel but with these they were able to build a slab hut. Later they built a woolshed and covered the roof with sail cloth for the first shearing.

Scott decided to leave after about six years and Gray and Marr bought his share between them. This included a large number of cattle and Gray, in an early form of asset stripping, overlanded them to Adelaide and sold them to recover some of the money they had paid to Scott.

Marr continued to run Nareeb whilst Gray spent much of his time dealing in cattle throughout South Australia and Victoria. Then in 1850 they decided to split up. They

Timour, the chief of the local Aboriginal tribe. He died an old man and was buried on Nareeb Nareeb. From a pen and ink drawing by Elizabeth Gray, the wife of the founder of Nareeb.

divided the sheep equally between them and Marr received the cattle whilst Gray took over the property.

Still living in the old slab hut near the creek, Gray started to make improvements. But after only a few months fire roared through the country, killing most of his ewes and leaving the property devastated. There was not enough grass left to fill his hat.

He started to build another cottage but, because of the fire, the nearest timber was 30 kilometres away. Using his team of bullocks, and helped by his ex-convict driver, he brought in enough timber to build a small house of four rooms. By the time he had fenced it and made a flower garden he thought it 'looked rather well for those times'.

N

To Glenthompson

1·6 Km.

1·6 Km.

Red Gum

To Hamilton

HOMESTEAD

5 Km.

To Caramut

To Chatsworth

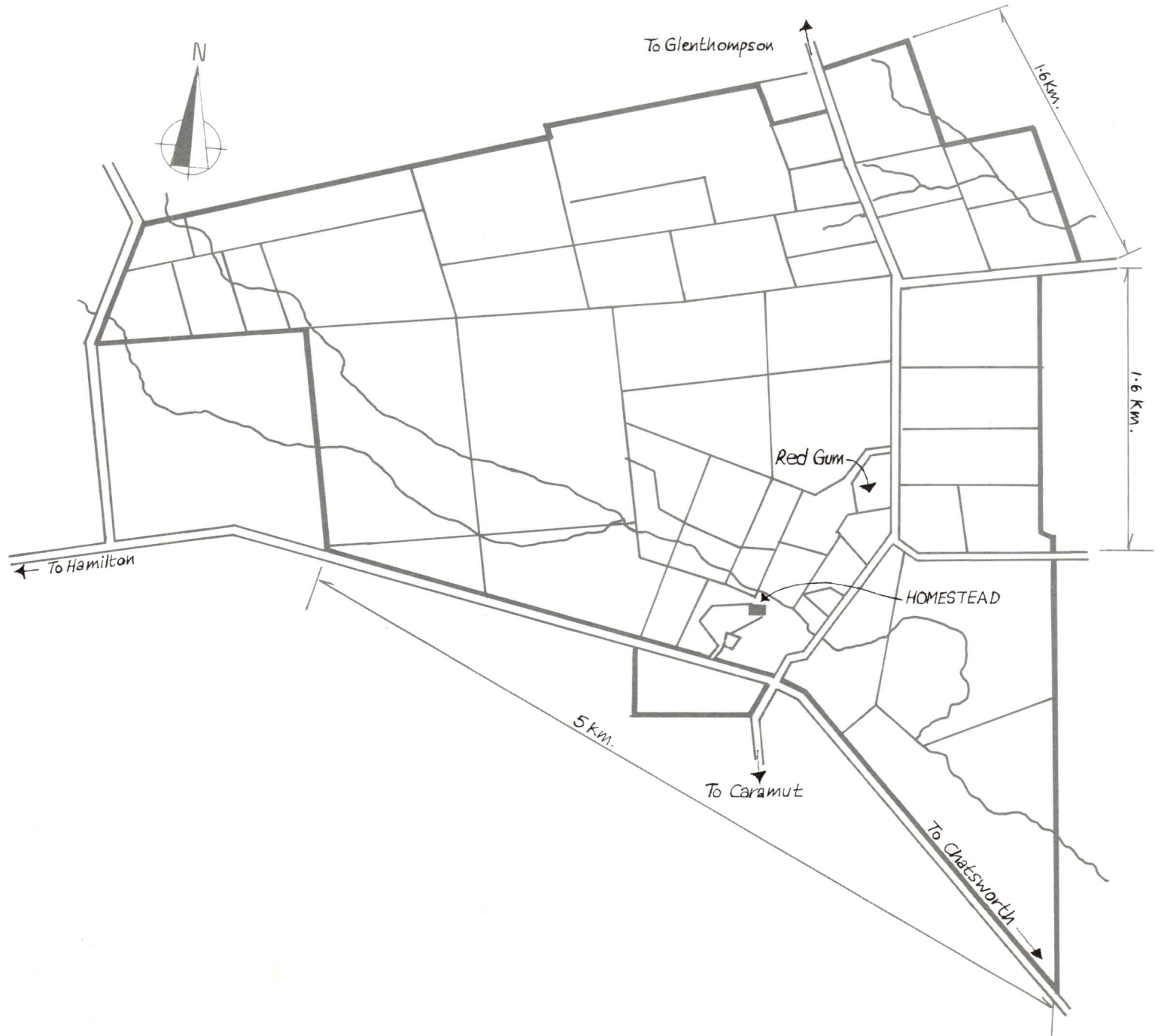

He also restocked the property and soon had a good flock of about 10,000 sheep. By the time he married in 1857 the place had fully recovered from the fire and his flock was one of the best in the district. Between 1864 and 1886 he won thirty-five first prizes, eighteen seconds and eleven thirds at the Hamilton Show, together with one gold and thirteen silver medals. In 1873 he won the Grand Championship for the best ram in a competition at Hamilton that was open to all the colonies.

By now a man of substance, he went overseas in 1876 but he returned alone, his wife and three daughters preferring to stay in England. They returned in 1881 but even though he had built a big new homestead for them they stayed only a year. Now nearly eighty years old, and with no son to follow him, he decided to sell Nareeb and join his family in England. He left Nareeb with regret but was, he said,

'relieved to be free from the annoyance now being caused by sheep shearers and others'.

The property was bought in 1886 and in 1905 it was sold again, this time to Geroge Maslin. It was he, in 1909, who sold it to Robert and Theodore Beggs, sons of a pastoral family who owned a large property at nearby Beaufort called Eurambeen. When this partnership broke up in 1913 Nareeb was taken over by another member of the family, Hugh N. Beggs. He sold off the sheep and restocked with sheep from Eurambeen, most of which were descended from fine-woolled sheep which had been imported from Tasmania in 1840. In 1922 he sent some of his ewes to be crossed with a ram from Uardry which had been recently bought by the neighbouring stud at Chatsworth House.

He believed that successful breeding required skill and patience and that when a type had been fixed it should be

Hugh Beggs, manager of Nareeb Nareeb.

Bill, the seventy-one-year-old supervisor who has worked at Nareeb Nareeb all his life.

Sandford Beggs examining a fleece in the woolshed at Nareeb Nareeb.

The homestead at Nareeb Nareeb, built after the previous one had been destroyed in the disastrous bush fire of 1940.

developed by matching potent sires with the very best ewes. And he was successful, for his fine-woolled sheep won twenty-nine Championships and five Grand Championships.

His son, Sandford Beggs, had been born in 1907. Educated at the Church of England Grammar School in Corio, he had since run a property called Meringa which had been bought for him by his father. Sandford Beggs married in 1935 and in the next few years, whilst continuing to live at Meringa, he took an increasing role in the running of Nareeb.

On the morning of Wednesday, 13 March 1940, the peace of Nareeb was shattered by a railway locomotive. Not that anybody there saw it, or even heard it. Several kilometres away, near Dunkeld, it threw out a spark which set fire to the grass alongside the line. It was a hot day and a strong north-westerly wind spread the small fire and quickly carried it across the dry ground. Soon it was burning on a broad front, destroying the country as it headed for Nareeb.

It reached there just after lunchtime and by then the fire was so big that it was impossible to fight it. The sky was overcast with smoke which was carried ahead of the flames by the wind. On the ground the fire swept across the paddocks, burning fences and trapping livestock as it roared towards the homestead.

A station hand went to the timber and concrete house to tell old Mrs Beggs that it was time to leave. She was unimpressed, thinking that the fire could not be as bad as he said. In any case, she was looking after a friend's parrot and she could not leave that behind.

Desperately the man caught the parrot and stuffed it into a large tin can before finally convincing her that the house was about to burn. But once outside the old lady turned back into the house again, even though the huge hedge behind the house was already well alight. She reappeared a few minutes later, clutching the stud books in one hand and swinging a silver teapot which she had noticed on the way out in the other. With the whole of the country now on fire, they drove to a big dam not far from the homestead and waded into the water.

Meanwhile, Beryl Beggs jumped into her father's Packard and drove at 100 km/h to Nareeb West State School. There

The dam near the homestead which was used as a refuge by family and staff during the 1940 bush fire.

she bundled the children into the car and drove like fury to the dam. She pushed the children into the water as one of the station hands drove to the water's edge and leapt in too.

Hardly able to see because of the smoke, they stood in the water and watched as the flames burnt the grass to the edge of the dam. The only humour came from the parrot, which continued to talk to everybody from inside the tin can.

When it was over, the whole of the property had been burnt out. The homestead was destroyed and only one of the many station cottages was still standing. Apart from that, there was simply nothing left.

Sandford Beggs, now desperately needed, was away on army maneouvres near Geelong and according to the army could not be contacted. It was only after his wife used some crisp language on a general that the news of the fire was passed to him.

He returned the following Sunday to a scene of total devastation. There were no fences left and the stock that had survived was now wandering all over the country. He spent the next three weeks recovering them and shooting those that had been too badly burnt to move. In one

paddock he saw a group of sheep camped under a tree and left them there whilst he dealt with the rest. When he finally went up to them he found that they were all dead, suffocated by the smoke.

When Sandford Beggs finally took stock he discovered that they had lost 12,000 sheep and that only 13 breeding rams were still alive. And he started to build the Nareeb stud all over again. It never occurred to him to do anything else.

He rebuilt the homestead and then built a new woolshed to replace the blackened remains of the old one and when his father died in 1941 Sandford and his wife moved on to Nareeb. There was another fire in 1944, but this time it swept through the top paddocks and did not threaten the new buildings.

By the time Sandford's son Hugh started work at Nareeb in 1961 the property was economically sound again. Hugh had been born at Meringa in 1938 and after going to school at Geelong Grammar he had taken a degree in rural science at the University of New England. He joined Nareeb as a station hand and then worked as a jackeroo before eventually taking over as manager.

Although it might seem that he had little choice in his career, he had always been free to make his own decisions about his future. If he was on Nareeb it was because he wanted to be, because it was an environment he knew and loved.

By the early 1970s he and his father were beginning to have doubts about the future of fine wool. The measurement of wool fibres now showed that a lot of wool that would have been classed as fine was actually not as fine as everybody had thought. With the use of scientific measurement likely to increase in the future, it seemed that the value of their clip would probably drop. Perhaps it was time for a change.

Whilst the big Peppin merinos were ideal for the arid parts of New South Wales and Queensland, they had never been as successful in the higher rainfall area of the Western District. So now the Beggs decided to breed a 'big northern sheep from the south', a sheep that would have the constitution of a Peppin with typical Western District wool.

The opportuntiy came in 1975 when the famous Chatsworth House stud was offered for sale. It had been based entirely on Peppin merinos from Wanganella and other Riverina studs and the Beggs thought it had just the blood they needed for their new type of sheep. In any case, it was too good a stud to see broken up.

The Chatsworth House stud brought them about 9,000 sheep and they were able to apply an extremely high culling rate as they combined them with their own Saxons. Within four years they had produced their big sheep, and although development is still going on they now sell sheep that produce heavy fleeces of sound wool and which are also suitable for the export trade to the Middle East.

In 1977 Nareeb Nareeb was burnt out yet again. This time the fire suddenly changed direction as it approached the homestead. The change saved the buildings but the fire overran 2,500 sheep which otherwise would have survived. Instead, they all died.

Nareeb Nareeb now runs about 14,000 sheep. They sell 1,000 rams each year and produce about 200 bales of wool with a permanent staff of five.

One of them, overseer Bill, is seventy-one years old and has worked on the property all his life. He joined up during the war but they sent him back as the stud was more important. He was angry about it and Sandford Beggs gave him the lease of a paddock so that he could run a few sheep of his own. From this, Bill was able to put his three children through college.

He climbs into the ute awkwardly because of arthritis in his hip. But as he drives slowly away to move a mob of rams to one of the far paddocks it is obvious that he knows exactly what he is doing. He stops the sheep ahead of some rising ground and walks on to look for another mob being brought in by one of the hands. They are ewes and if they meet up with his rams there would be a sort of sexual Armageddon.

They are not in sight, so we go on. He knows every stone and if there was much grass on the paddocks he would probably know every blade. A ram breaks from the mob and

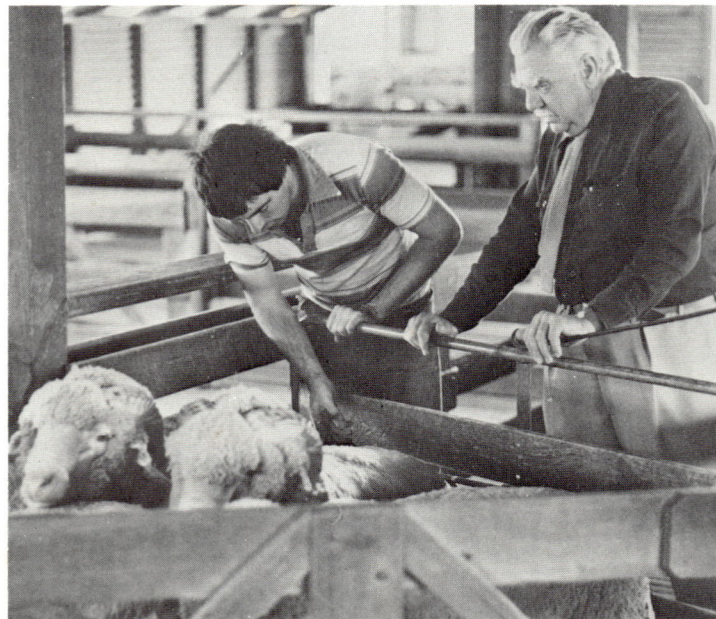

Sandford Beggs shows a group of rams to a client.

a dog chases it briefly before giving up. As it ambles back, the ram keeps running. Bill curses and swings the ute round to head it off. The dog's future looks anything but certain.

'Funny about dogs,' he says later. 'They get old too. Had one once that was a real beauty — just after the war it was. But I had to shout awful at him all the time. If I told him he was a good dog he'd slink off home. We had some refugee people helping us at that time, Poles I think, or maybe they were Lithuanians. Anyway, they'd seen a bit of the war.

'One day, after I'd told 'em what to do I got into the ute and looked round for the dog. He was a fair bit away so I yelled "'Ere, you bloody bastard, come 'ere and get in the bloody back." Pretty bloodcurdling it was, but I didn't think anything of it until I saw this young bloke, about nineteen he was, jump up on to the back of the ute and sit there shivering with fear. I felt real awful about it.'

Meanwhile Sandford Beggs, who is now seventy-five, has spent the morning driving round the paddocks with his wife to check the water points. She has been with him to open the gates and by the time they have finished she has counted twenty-seven, but she thinks she might have missed some.

They now live in a comfortable house called Redgum which is about a kilometre from the homestead. There are fishing rods on the wall and the shelves are full of books about trees and birds. He settles down to read the paper and the phone rings with an odd warbling sound. Mrs Beggs answers it and responds warmly.

'Are you feeling better, Malcolm?' she asks, her concern genuine.

Sandford Beggs looks up and tells his wife to send his best wishes, but she does not hear him. He goes back to his paper.

If you have spent part of your life rebuilding Nareeb Nareeb with thirteen rams and no fences, having a son-in-law who was until recently the Australian Prime Minister might not seem all that important.

LANSDOWNE

In the summer it is unusual to see clouds in this part of western Queensland. Day after day, the sky is a deep blue canopy that stretches without interruption right down to the horizon. The sun throws deep shadows and the temperature climbs steadily throughout the day. At night countless stars twinkle as the heat of the day leaves the ground. But not all the heat dissipates and even before the sun rises the air is still soft and warm.

But today is different. As the sun climbs into the sky a few clouds gather above the western horizon and by the afternoon they are dotted half way across the sky. They are fleecy-white and quite small, but the men at Lansdowne look at them often, hoping they will build into a storm that will drop rain on the dry paddocks. They have had virtually no rain for six months and not very much in the year before that.

Now, a tropical cyclone is meandering a few kilometres off the coast at Rockhampton and these clouds might be part of it. It is a slim chance, to be sure. Lansdowne is nearly 500 kilometres inland from Rockhampton and nobody expects the rain to carry that far. On the other hand, the clouds might be part of a different weather system that has nothing to do with the cyclone. The technicalities don't matter very much — nobody cares where they have come from provided they drop some rain.

But they don't, and the day ends as dry as it began. The clouds are still there the next day and the day after that. Although they are still white, refusing to take on the shades of blue and grey that would indicate a storm, hope remains.

This part of Queensland was first explored by Thomas Mitchell, who in 1846 discovered lush country and an important river which he felt sure would run north-west into the Gulf of Carpentaria. He called it the Victoria. The following year his assistant proved that the river was actually a northern tributary of Cooper Creek and that instead of running into the Gulf it turned south-west to disappear into the waste land of Sturt's Stony Desert. It was hardly important enough to justify such a distinguished name as Victoria and in any case a river in the Northern Territory had already been given that name before Mitchell's discovery. So it was called the Barcoo River instead.

Europeans did not see this country again until 1858, when Gregory passed through it while searching for the missing

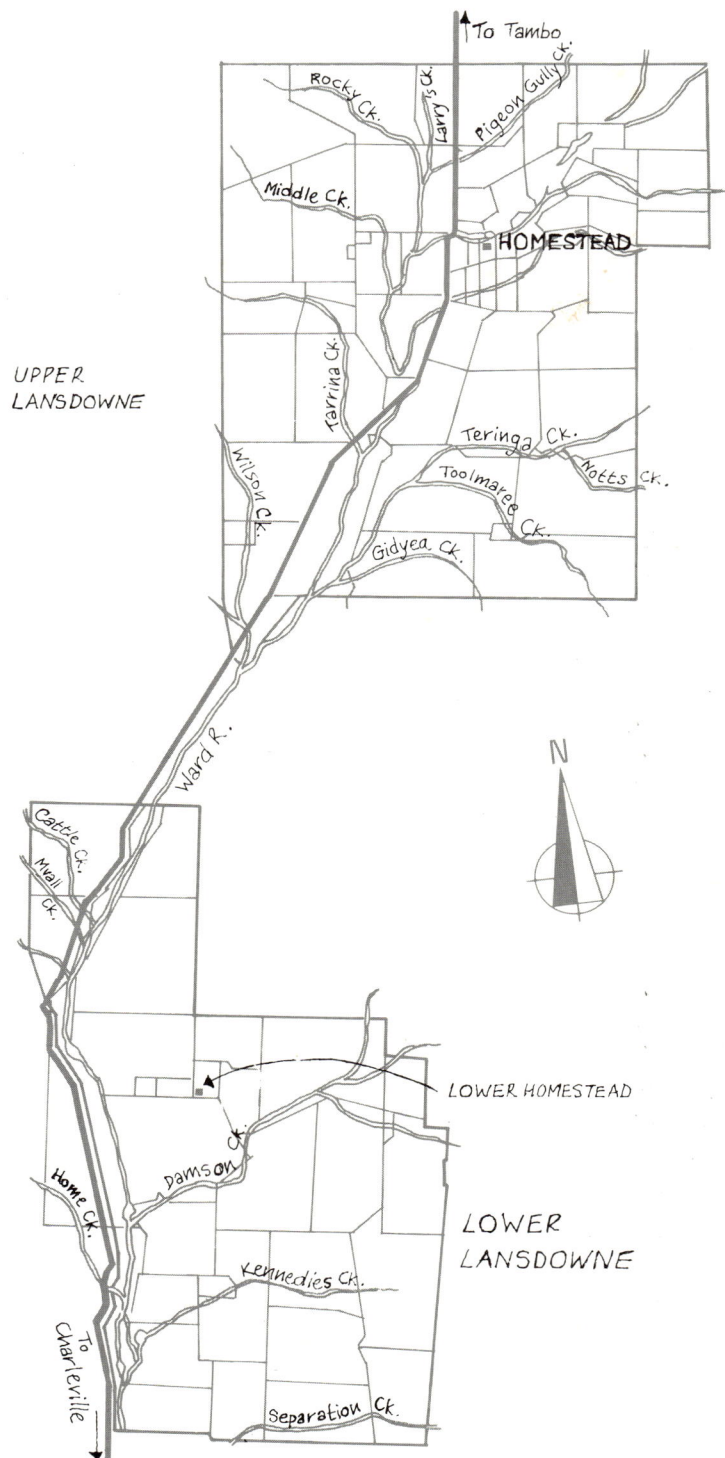

Ludwig Leichhardt. The country he saw was far from lush: all the vegetation had gone and the Barcoo River was dry. It was an early indication of what the seasons could do in this part of Queensland.

With other land more readily available, the country round the Barcoo attracted little interest. It was not until 1861 that the first white settlement took place, when three men, Burne, Mayne and Ward, applied for licences to occupy an area which they later called Lansdowne. They stocked it with sheep but, like many landowners at that time, they did little to improve it. They built a crude homestead, fenced some small paddocks nearby, and settled down to run their sheep.

Indeed, they needed few refinements. The run consisted of 431,040 acres and the sheep were free to roam during the day. At night they were herded into temporary pens made from wattle to protect them from attacking dingoes. There was no breeding programme, no neighbours, and no need for fences.

In 1875 this huge property was taken over by George Fairbairn and things began to change. Fairbairn was an accomplished pastoralist and a few years earlier had brought John Turnbull out from Scotland to manage another property of his, Peak Downs. He now took Turnbull into partnership and gave him the job of running Lansdowne from his base at Peak Downs.

Five years later Fairbairn offered to sell his share of Lansdowne, together with another property of his called Evesham, to Turnbull. Encouraged by Arthur King, grandson of the third Governor of New South Wales, Turnbull agreed. But it was an expensive deal and far beyond his personal means, so with the help of King, Turnbull formed a company to raise the money. It was called the Lansdowne Pastoral Company and by the time it took over the two properties Turnbull owned 25 per cent of it and moved on to Lansdowne as Chief Station Manager.

By this time Lansdowne had improved considerably from the rudimentary run of its first owners. The homestead was now a four-roomed building of sawn cypress slabs and had a separate kitchen nearby. Even more impressive was the seventy-stand woolshed which had been built from pines brought from the other side of Tambo.

The first two years were particularly good and by 1884 there were 304,000 sheep on Lansdowne, together with several hundred horses and cattle. But then the seasons changed and an unusual sequence of drought and flood showed the folly of overstocking the country. In 1885 there were only 152,000 sheep on the property (many had been moved to Evesham) and Turnbull generously offered to take a reduction in salary. The Board, equally generously, agreed.

The lessons were well learnt, however, and as the seasons improved stock numbers were kept low. It was always tempting to run large numbers of stock when the paddocks were lush with feed and the waterholes full. But these conditions could change very quickly and when they did there was little that could be done to keep sheep alive, and

Tom Kinross, manager of Lansdowne.

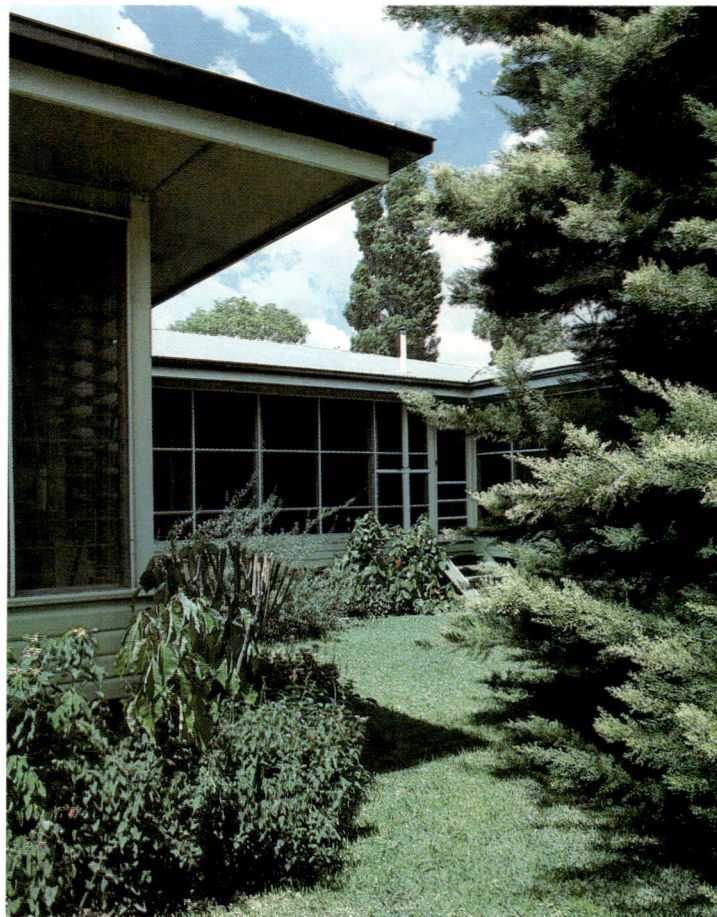

Part of the Lansdowne homestead.

A jackeroo unloads his horse for an early morning muster.

Moving a mob across one of the parched paddocks at Lansdowne.

almost no hope of selling them. Paddocks would then be severely overgrazed and it could take years of good seasons for them to recover.

Turnbull retired in 1889, although he retained his title of Chief Station Manager and remained on the Board. The management of Lansdowne was taken over by William Young, an intensely active man who was tough, thorough and competent. He expected his men to be the same and the working day at Lansdowne started well before dawn and ended after dusk. It was known, sometimes wearily, as a 'Billy Young day'.

But enthusiasm and long hours could not solve all his problems. He was soon caught up in the bitterness of the shearers' strike — a bitterness that was particularly marked in this part of Queensland where the threat of violence was never far away. This was followed by a period of below-average rainfall which slowly became a drought. When Young resigned in 1898 there were less than 100,000 sheep on Lansdowne and the company was barely paying its way.

Stock numbers fell even lower as the drought continued and by 1902 they had less than 40,000 sheep. The following year they produced only 700 bales of wool, whereas twenty years earlier they had sent away more than 2,000. It is hardly surprising that managers did not stay long. Young's successor stayed only five years and the property was in drought for the whole of that time.

As conditions slowly improved, so the stock increased and by 1908 they had over 120,000 sheep. The procession of managers also came to an end when Cecil Mills took over in 1924. A man of considerable experience, he had been overseer at Haddon Rig and had worked on a wide range of properties. His experience was soon put to the test, for his early years at Lansdowne were far from placid. In his first year a huge fire burnt out much of the property and the following year the rain failed.

There were about 100,000 sheep on Lansdowne at that time and Mills kept them alive by hand feeding. This was the first time it had been attempted on such a large scale in this part of Queensland, and indeed has seldom been attempted since. The feed, mostly maize and lucerne, was taken into the paddocks every day by truck. It took 113 grams of maize each day to keep one sheep alive and the cost of providing it on such a large scale was considerable. Fortunately the drought did not last long and when it ended Lansdowne still had most of its sheep.

The Depression dominated the Thirties, but the start of the Second World War eventually brought better prices and stability back to Lansdowne. In 1943 Mills retired because of poor health and Graham Lilley took over as manager. By this time the property was badly in need of attention but with materials and manpower both in short supply there was little chance of making improvements. Indeed, Lansdowne could only function at all because there were a few old hands who could still do most of the work.

It was not until 1950 that work could start on renovating the property. A new homestead was built, cottages and sheds

A management meeting at the paddock fence.

were replaced and improvements made to water storage and supply. That year also saw one of the highest rainfalls ever recorded at Lansdowne — 37 inches — and although this was in marked contrast to the succession of dry years in the past it also resulted in fly strike on an unprecedented scale.

Nevertheless, in 1950 a bale of Lansdowne wool was fetching about £173 and the total clip of 1,686 bales produced an incredible income of £291,678. The following year the company paid a dividend of 65 per cent. The wool boom was beyond anybody's dreams but, like most booms, it was short lived. By 1954 the clip was down to 957 bales, the price per bale and had dropped to £108, and the income had fallen by nearly two-thirds.

The following year the United Graziers' Association of Queensland proposed a reduction of 15 per cent in the rate paid to shearers and, not surprisingly, all the old bitterness came to the surface once more. When the Industrial Court granted a reduction of 10 per cent, the shearers went on strike.

Lilley had anticipated the trouble and had managed to shear 3,000 young rams before the strike started. The following May he brought in a team of New Zealand shearers to shear the stud ewes and 700 bales were kept under armed guard in the woolshed until they could be moved. The

An old steam boiler which was used to heat water at the washpool.

along the watercourses but the paddocks are vast open spaces with nothing much above knee height. If your eyes are good enough, you can just see the fence line in the far distance.

The property is watered by fifty-five dams, four artesian bores and three sub bores and gets about 20 inches of rain a year. It carries 55,000 medium-woolled Haddon Rig Peppins on its red and black soil and they produce about 1,500 bales of wool a year. They sell about 700 rams each year at Lansdowne and most of them go to similar properties in western Queensland.

Tom Kinross is the present manager of Lansdowne. He was born at Inverell in New South Wales in 1927 and was educated at The King's School in Parramatta. Even then he knew he wanted to work on the land and in 1947 he went as a jackeroo to a property near Moree before moving to south-west Queensland. In 1951 he got his first job as an overseer at Minnie Downs in Queensland and returned there as manager in 1956. He stayed there for twenty years and when the property was sold in 1976 he was asked to take over as manager of Lansdowne.

At that time the property was in good order but a run of bad seasons had left the stock in poor condition. With sheep classer Dick Jago, Tom Kinross concentrated on increasing cutting weight whilst retaining a fibre diameter of about 21–23 microns. The classing is done by visual inspection, with objective measurements used to confirm the selection. The poll flock has been increased so that it is now as big as the horned flock. Demand is high and he expects it to be even higher in the future.

As an experienced and professional manager, Tom Kinross has no difficulty working for a company which in turn recognises his expertise and gives it full rein. They work out budgets together and any significant proposals, such as a change in the stocking rate, are discussed at the two meetings which are held on the property each year. The company's pastoral director, Richard Turnbull, is a direct descendant of the founder of the Lansdowne Pastoral Company and he has a property of his own not far from Lansdowne.

There is not much scope for pasture improvement at Lansdowne but the use of non-protein feed supplements seems promising. Although droughts are not all that common there is often a shortage in the feed. Tom Kinross is a keen follower of technology and hopes that science will soon solve the problem of fly and parasite control. Both account for a considerable amount of time on Lansdowne and more effective control would do much to simplify management of the stock.

Tom Kinross lives with his wife and two children in the homestead at Upper Lansdowne, which is the northern section of the property. It is 20 kilometres from the small town of Tambo, which sits quietly on the road between Charleville and Blackall. There is no air service, but every morning at 7.35 the Greyhound bus pulls in on its way from Brisbane to Longreach. It has left Brisbane at 6.30 the previous evening and now, thirteen hours later, it stops at

following August, with the strike still on, he was faced with the job of shearing some 35,000 flock sheep. In the end it was done by only two shearers, one Australian and the other a Maori, helped by four station hands and an occasional jackeroo. They finished in October, just as the ten-month strike came to an end.

The Lansdowne Pastoral Company Ltd of Melbourne still owns this property, although it no longer owns Evesham. Lansdowne has operated as a stud since 1881, when it was founded with rams brought from Tasmania. These were soon replaced and most of the later imports came from Haddon Rig. A poll stud was started in 1945 with one ram that had turned up in the horned stud.

The property now consists of 153,715 acres split into two sections whose boundaries are about 8 kilometres apart. It is high country, some 426 metres above sea level, and stands on a watershed. A few kilometres away at Tambo the Barcoo River runs to the north-west before curving away to join Cooper Creek. At Lansdowne, though, the land falls the other way and the Ward River, which runs through the property, flows south to join the Darling.

The open downs country rolls gently away to the distant horizon. There are no hills and the watercourses are simply the hollows of the undulating country. There are a few trees

swimming in the dams, provided you don't mind coming face to face with a few yabbies.

The rest of the staff will spend the morning lamb marking and they are already loading horses into the trailer. Soon they are driving past the big woolshed and heading across the gently undulating country to a far paddock.

The first job is to muster the ewes and lambs and drive them to a set of yards in another paddock. The ute stops, horses are unloaded, and within minutes the muster has begun. Working the sheep into the wind, the men soon have a mob of several hundred and move them steadily to a gate in the top fence. The team work is effortless. Dogs keep the sheep from breaking from the mob and the men ride easily behind them. With Ian driving the ute slowly behind the riders, they walk the mob quietly through the paddocks and into the yards.

They unload the ute and assemble the equipment. They stand a tubular steel frame against the outside fence of the yard. Its top is no higher than the fence and consists of a line of cradles which will hold the lambs on their backs with their hind legs in the air.

Two jackeroos go into the yards. They each catch a lamb and lift it over the fence to secure it face up in the cradle. Before all the cradles are full, Ian and others have already started work. Whilst one man notches the ears, Ian lifts the tail of the lamb and feels for the fourth joint. Deftly, with a single movement, he cuts the tail at that point and throws it on the ground. If the tail is not docked it soon becomes stained and a fly strike is then almost inevitable. Fly strikes are extremely painful to sheep and this is an essential way of avoiding them. If the lamb is a ram, he also removes the testicles just as quickly before going on to the next. Another man holds an upturned can of antiseptic over the tail, then follows Ian down the line.

They work quickly. When they have finished the cradle is tilted forward and all the lambs fall into the paddock. They land on their feet because the cradle is designed to make sure they do — if they landed on their rumps there would be a high risk of infection.

The lambs walk away. Some go off into the paddock whilst others go to the fence and bleat at the ewes still in the yards. They are bleeding from the stump of the tail but their bleating seems more of outrage and discomfort rather than pain, as Ian and the rest of of the men are very skilled. The work on each lamb takes no more than ten seconds.

The work goes on and the pile of tails grows. When there are no more lambs in the yard the ewes are counted out into the paddock and the yard refilled with more lambs and ewes. It goes on for hours, but the pace never slackens and the movements of the men are as deft as before. Billy Young, when manager of Lansdowne, is said to have marked 9,000 lambs in one day in a paddock not far from here. It is thought to be a world record, but nobody knows for sure.

When the work is over they put matches to crumpled cigarettes and add up the number of ewes that they have counted out through the gate. They then count the tails they

Daphne Kinross using a pair of shearing blades to prune the homestead garden.

Tambo for breakfast. Soon it will be off again and for most passengers Tambo will be no more memorable than any of the other small towns on the way.

But Tambo is important to the people of Lansdowne. It provides a social life outside the station, as well as a primary school for their children. Even so, it is an isolated place and the biggest town beyond Tambo, Charleville, is over 200 kilometres further south.

A few days later the clouds are still there, although they still look harmless. It is early morning and vehicles arrive in the station square and swirl the dust as they come to a halt. Ian Sutherland, the overseer at Lansdowne, is organising his men for the day. He has a staff of eleven, including jackeroos, but two are mechanics who spend most of their time in the maintenance area.

He details two of the jackeroos to spend the day checking water levels in the dams. They will drive to each dam and one of them will swim into the middle and then dive to the bottom to get some idea of how deep it is. They are quite pleased with the prospect. Clouds or not, the day is going to be very hot and there are worse ways of spending it than

The woolshed at Lansdowne. Manager Graham Lilley kept 700 bales of wool under armed guard in this shed during the 1956 shearers' strike.

have thrown on to the ground to find the number of lambs. Simple arithmetic gives them a percentage of ewes that have produced lambs.

This figure, the lambing percentage, has much significance on a sheep station as it is a measure of the fertility of the ewes and the future size of the flock. If there are a large number of twins born, lambing can be more than 100 per cent, but the lambing percentage is usually about 80 or 90. Today the figure is about 60 per cent and this is a result of the drought and its effect on the feed in the paddock.

As they start to load the equipment on to the ute, the men look at the clouds again. They are much bigger now and have turned darker. The wind is in the right direction and if it stays there there might be rain later. They talk about it as they mount their horses to move the lambs and ewes back to their paddock.

Later that day I am driving to Lower Lansdowne with Tom Kinross. Although the boundaries of the properties are only 8 kilometres apart, the two homesteads are about 32 kilometres apart. By now the clouds are huge, stacked above each other in great masses of purple and grey. It is going to rain.

It starts before we get to the first boundary. Gentle at first, small spots drop on the windscreen to mingle with months of dirt and dust. Then it is raining hard and drumming noisily on the metal roof. And then torrential, blinding rain blots out everything. The wipers cannot cope and we can barely see the front of the car. Crawling along, we throw up great clods of mud from beneath the car and we can hear the tyres swishing through water that is already centimetres deep. As the car bucks on the uneven road lashings of water are thrown back over the bonnet. It is more like being at sea than driving through drought-stricken Lansdowne.

We push on, the rear of the car sliding in the mud. Tom spins the wheel to keep it on the flooding road, talking cheerfully because it is raining. We might have to turn back, but we are on black soil and we would bog immediately if we left the track. By the time we reach a patch of firmer red soil the rain is easing and we decide to go on. More than an inch has fallen in less than an hour.

Tom drops off at the woolshed at Lower Lansdowne and the sun is shining again as I drive off to look at the paddocks. The ground is damp but this part of the property has missed the centre of the storm. I drive slowly — but not slowly enough. The track runs through a small cutting and as I round a bend I am suddenly surrounded by fast-running water. I am almost through it when the wheels lose their grip on the mud below and the car slides to a standstill. I select four-wheel drive and use gentle power to ease the car backwards and forwards, but I can't break the suction of the mud. Behind the vehicle the water swirls in a creek that has been running for only a few minutes.

Ian Sutherland, the Lansdowne overseer, lamb marking in one of the paddocks. If a lamb's tail is not docked, it soon becomes permanently stained and a fly strike is almost inevitable.

The car is down to the axles in mud and even if I can get it out I could not be sure of crossing the creek to get back. But I cannot get it out and I need help. I reach for the radio.

'Any Lower mobile, do you read, over?'

'Yeah, got you. Trouble, over?'

'I'm bogged in East Damson and the Damson Creek is running. Over.'

'Be there in a few minutes. Out.'

Soon a bigger four-wheel drive comes round the bend in the track and stops on the other side of the water. A big man gets out.

'Shouldn't be any problem, but it would be in half an hour.'

We run a steel cable between the two vehicles. I get into mine, start the engine and hold it in reverse, slipping the clutch until I feel the pull of the cable. I ease the clutch and

the vehicle moves slowly out of the mud as it is pulled backwards through the creek. It is obvious he has done it many times before. He coils up the cable and says he might need it again soon if the creeks are running that quickly.

As I drive back to the woolshed the ground is no wetter than it was before. I pick up Tom and tell him about the creek. It makes him more cheerful than ever.

'Look,' he says, 'you'd better drop me off at the homestead for a minute. The wife here is in town and expecting a baby any tick. Shouldn't be any problems but I'd better find out what's happening.'

Ten minutes later he comes out of the house and I can see immediately that the news is bad. I wonder how to break the dreadful silence.

'So, what's happened?'

'They only had twenty points here,' he says.

But it still looks like the end of the drought.

THE JACKEROO

A jackeroo is a young man (or, less frequently, a woman, who is then known as a jilleroo) who works on a property for a low wage in return for tuition. He is a cross between an apprentice and an officer cadet and it is assumed that one day he will run a property of his own.

Because of this, jackeroos have traditionally come from property-owning families. For them, it is a way of acquiring more varied experience and, often, more expert tuition than they would get at home, and so make them competent to take over the running of the property when the time comes. Recently, however, more and more jackeroos have come from city backgrounds after deciding on a career in property management. Their aim is to become professional managers of properties owned either by pastoral companies or private owners who might not have the time or expertise to run them themselves.

In the sheep industry it is rare to find anyone running a big property, even as the owner, who has not been a jackeroo. For a manager of a company-owned property it is an almost essential qualification, although often the only one.

The term of this 'apprenticeship' is usually four years, although it need not all be spent on the same property. During that time he will (or should) be taught all aspects of animal care and property management. At first he will spend most of his time working under supervision alongside the station hands, doing the same jobs but for less pay. Later, however, he will become more involved in specific aspects of management such as breeding techniques, pasture improvement and financial control.

In the cattle industry the system is less formal, usually shorter, and is designed initially to produce good stockmen who might then go on to management if they have the ambition and ability.

A big property might have a dozen jackeroos in various stages of training at any one time and they will live together in their own quarters. A small property might have only one and then he will probably live in the homestead with the owner's family.

Vic Moar is nineteen years old and is one of three jackeroos on a sheep and cattle property in northern New South Wales. He was born in Sydney but after a

One of the jackeroos at Haddon Rig.

number of family moves he finished his education at Armidale High School. His father supplies fuel to a number of nearby properties and the owner of one of them invited Vic to work for him for a couple of weeks after he left school. They got on well together and at the end of that time Vic was invited to stay as a jackeroo.

Today he was up at 6 o'clock and his first job was to catch his horse. He then went into the paddocks to check the cows that were calving and returned at 7.30. He then milked the domestic dairy cows, which took two hours, and broke for morning smoko of tea and cake.

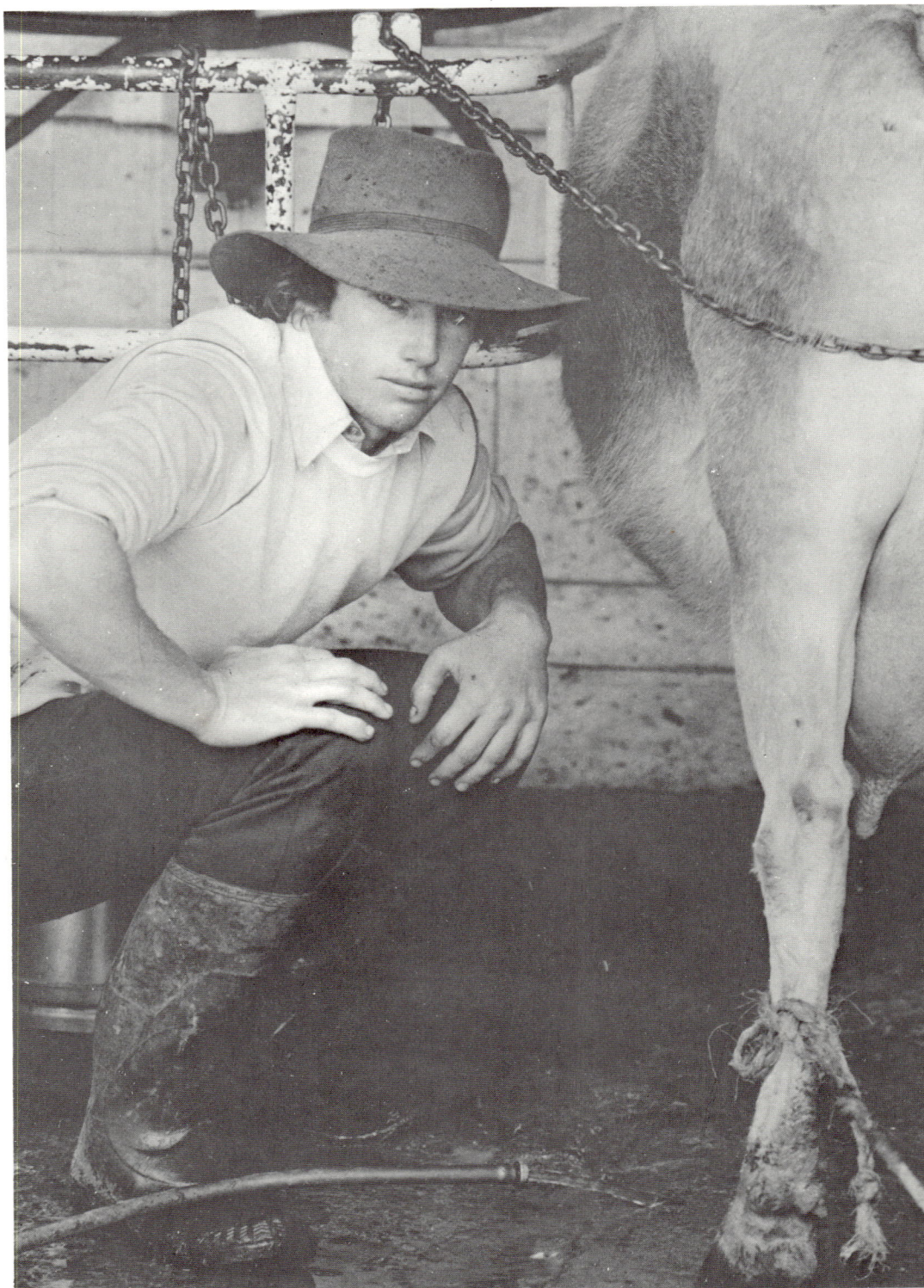

Vic Moar. Born in Sydney, he is now one of three jackeroos on a sheep and cattle property in northern New South Wales.

At 10.30 he was sent to the woolshed and shore ten sheep that had been missed during the recent shearing, then did general work in the yards until lunch at 12.30.

He was back at the yards an hour later and with two others started to drench sheep. They did 1,800 (although he did the last 300 by himself) and finished about 5.30. He then separated the milking cows from their calves for the night, fed them, then loaded the ute with bales of hay and drove round to drop them off at five feeding points. He finished work at 6.15, after a twelve-hour day, and was in bed by 9 o'clock.

He works five days a week and spends the weekends at home unless he is needed on the property. He is paid $115 a week after tax and from that he pays $40 a week in board.

'I love it,' he says. 'Hardly a day goes by when you don't learn something. It is always what I wanted to do, but I never thought I would have the chance.'

ANAMA

An art critic once said, with that mixture of acidity and accuracy that makes such comments memorable, that there were only two types of landscape paintings in Australia. One had gum trees with a flock of sheep between them, and the other had a flock of sheep with gum trees at either end. And if this is true, they could all have been painted at Anama.

The country of Anama, near Clare in South Australia, is gently undulating and, now the drought is over, refreshingly green. It is lightly timbered with trees grouped in copses in the paddocks or making straight lines along fences and roads, to give shade from the sun and protection from the wind. Blue gums and she-oaks, huge in their maturity, add substance and scale to the undulating landscape as it stretches away to the distant hills.

The sunlight, chasing shadows in the wind, brings light to the landscape. At first it is on the distant hills then, quickly and more dramatically, it moves nearer to sweep across the clusters of sheep as they wander, heads lowered, across the dappled greenness of the paddocks. It is a rural scene which, although commonplace in distant places and earlier times, inspired sonnets and symphonies as well as paintings.

For a moment those places and times seem not so far away, then a flock of galahs wheels in alternate pink and grey against the blue of the sky. The trees become eucalypts again and the lake, which seemed so perfectly placed by a landscaper of the noble school, becomes an Australian dam once more.

But this glowing scene loses nothing in the transition, for it now personifies Australia. It is the living reality of those countless paintings which, hanging in suburban houses, remind others of this Australia beyond the cities. And it would not be complete without the gum trees and the sheep. Sheep have always been important in South Australia because they made it possible to settle the arid areas that make up much of the State. But droughts, poor natural pastures and the shortage of surface water demanded a merino that was big and hardy and of good 'constitution', so the result was a different strain of merino: the South Australian.

Whilst this merino took a long time to develop, the sheep population nevertheless grew dramatically in the early years as pioneers pushed further away from the more settled areas. In 1837 there were only about 5,000 sheep in the whole of

George Hawker, who founded Bungaree in 1841. His sons divided the property in 1906 and Walter Hawker took the area known as Anama.

South Australia. Two years later there were more than 100,000 and in 1841 there were a quarter of a million.

It was in that year that George Hawker took up land near Clare at a place called Bungaree. He had arrived from England a few years earlier and had settled at Kapunda. There he lost the money he had brought with him and, in the best traditions of the day, he asked his father for more. He was given some, together with a firm declaration that this would be the last. He then joined an expedition to the unexplored north and it was on the return journey that he first saw the land at Bungaree.

The property he established stretched from Watervale in the south to the Broughton River in the north and covered half a million acres of good land. He bought 2,000 ewes descended from Macarthur's Camden flock from a property near Bathurst in New South Wales and overlanded them to

To Jamestown · Canowie · Hallet · To Peterborough · Mt Bryan East · N · Gulnare · To Port Augusta · Spalding · Booborowie · Mt Bryan · Caroona · Yacka · Andrews · Anama Booborowie · Leighton · BURRA · Redbanks · Anama Clare · Hilltown · Hanson · Worlds End Creek · Hart · CLARE · Blyth · To Adelaide · Farrell Flat · Seven Hill · Merildin · Black Springs · Emu Downs

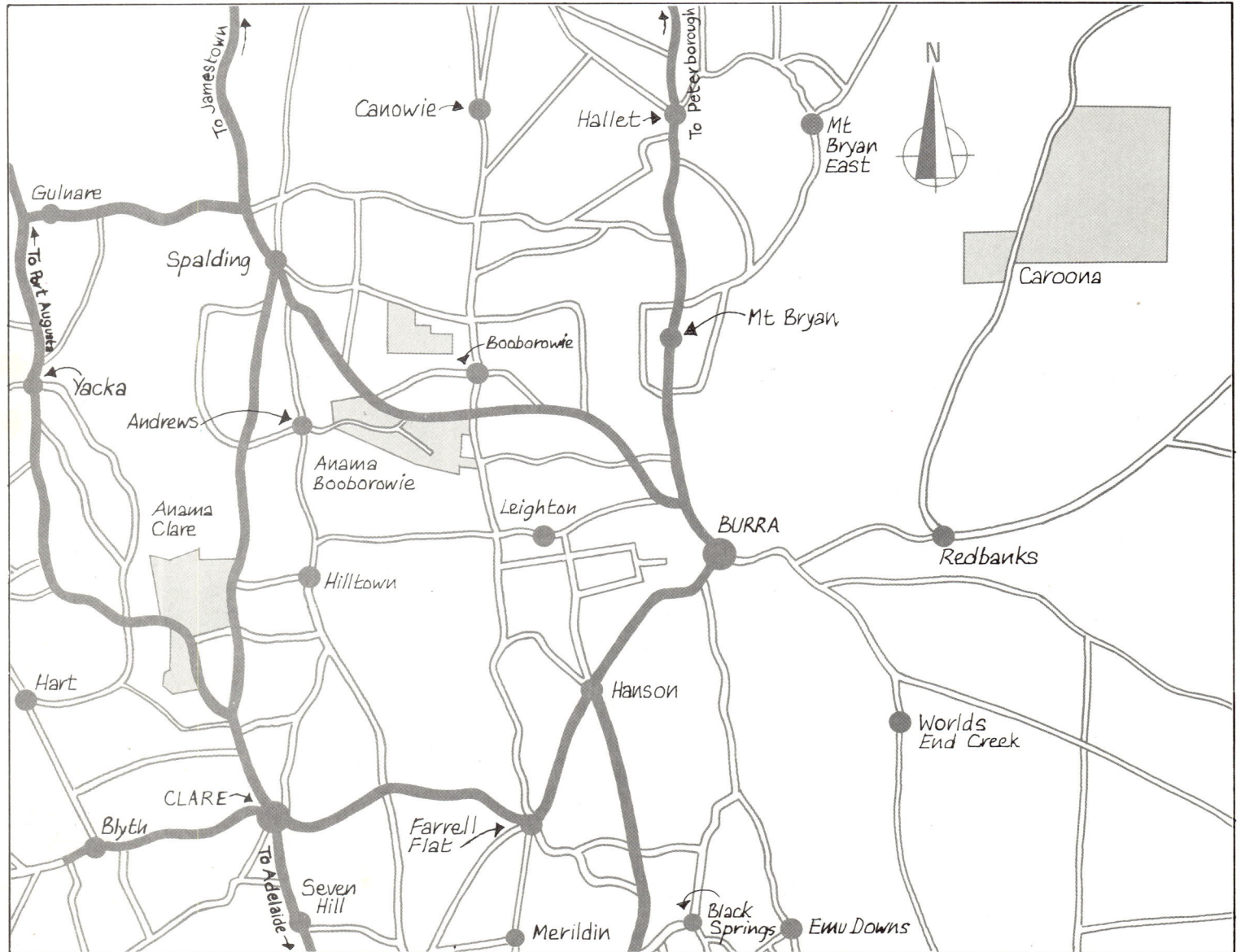

Bungaree. He also brought ten men to help him fight the Aborigines, for this country was more turbulent then than it is now. Indeed, when he married it is said that his wife was the most northerly white woman in the entire State.

His flock soon increased to more than 100,000 and it was looked after by fifty-two shepherds who lived a solitary life on the remoter parts of the run. And to look after the shepherds Hawker brought out a man from England called John Noble. Once a shepherd himself, Noble had an almost uncanny skill in counting sheep. He would hold the mob against a fence and then get four men to run them past him. When a hundred had passed he would call 'tally' to a man who recorded it, and when less than a hundred remained he would count them one by one. He could never explain how he knew a hundred had passed him, but he was never wrong.

In 1861 Hawker imported five or six Rambouillet rams from France and, although other rams were bought from Australian studs over the years, it was these Rambouillets that were the foundation of the Bungaree flock. Hawker aimed to produce a large-framed sheep and then to cover it with a fleece of long-stapled combing wool. It was a slow process, but by the time he died in 1895 Bungaree was

supplying South Australian merinos to many commercial flocks in the arid parts of the State and beyond.

John Noble continued to manage Bungaree until he retired in 1901 after spending forty-seven years there. Hawker's sons then took over the running of the property but in 1906, because of the threat of a tax on large landowners, they decided to divide the property between them and go their separate ways. The land was surveyed into equal parts and these were distributed to the brothers by drawing lots. It was Walter Hawker who drew the land to the north which was called Anama.

By 1910 Walter Hawker had a flock of nearly 2,500 sheep at Anama and they were all of pure Bungaree descent. Then three years later he bought a ram from Wanganella for 1,700 guineas and also leased its sire for six weeks for about the same price. The new blood was not a success, however, and in 1920 Walter's son John discarded the Wanganella influence and reverted to the original Bungaree sheep to develop what was to become the Anama merino.

The only rams he brought in from outside came from studs owned by other descendants of the Hawker brothers or from other related studs in South Australia and during the next

fifty years he developed the Anama merino for strength and constitution. Their horns were wide set, faces were soft and free of wool, and fleeces were long-stapled and strong.

But when John Hawker died in 1970 he left an industry that was changing more rapidly than at any other time in his life. Prices were low, economics were tough, and technology was already challenging the traditions of the past. Anama had to change too.

Today Anama consists of three separate properties and of these, only the 5,400 acres of undulating country at Clare was part of the original Bungaree. It is 14 kilometres north of Clare and most of it forms a wedge between the roads to Gladstone and Jamestown. There is, it seems, nothing harsh about this country. The average rainfall of 20 inches is generous enough in most years to provide an ample cover of feed which carpets its gentle hills. But that carpet is not always what it seems.

Some paddocks, especially on the western side, have been almost overrun by wild tulip which unfortunately is poisonous to sheep. Those that have grown up in these paddocks have learnt to ignore it, but sheep from paddocks that are free of it are often tempted by its novelty and losses can be considerable. Elsewhere, some of the gentle hillsides are covered with a rigid layer of stones. If the feed is high the stones are difficult to see, but they inhibit growth and make it difficult for sheep to move freely round the paddocks. Clearing them is a costly job which will go on for many years yet.

Ryves Hawker runs this part of Anama from the homestead which his father converted from a four-roomed share farmer's cottage whilst his own father lived in the now demolished homestead that he built in 1906. Light and very comfortable, the house now combines a pleasing blend of the traditional and the modern in much the same way that Ryves Hawker does.

Born in 1941, he was educated at St Peter's College in Adelaide and then went as a jackeroo to Belacres near Meningie. From there he went to Headingly, near Scone in New South Wales, and then in 1961 he went to the Royal Agricultural College at Cirencester in England. Gaining his Diploma of Agriculture in 1962, he worked as a relief manager on mixed properties in Kenya for nine months as a means of reaching South Africa to see the major studs there.

He spent three months touring sheep studs in Cape Province and South-West Africa, listening to their owners' breeding philosophies and seeing the different types of sheep they bred. He was happy there and when he was offered a manager's job in South Africa he was tempted to stay. But his father, having sent Ryves to England with £1,200 on the assumption that he would find his way back when it ran out, now recalled him.

Returning to Australia, Ryves spent nearly a year with his father on Anama and then went as an exchange jackeroo to Sturt Meadows in the very different country of Leonora in the north-eastern goldfields of Western Australia. There he

An original copy of a poster advertising the sale of land at South Booborowie in 1912. Anama still uses land which Walter Hawker bought at that sale.

learnt a great deal about running sheep in arid areas and benefited greatly from the knowledge of its owner, Geoff Chumley. 'The danger of being on a family property is that you can spend your life doing what your father did, both good and bad. You don't have the chance to question all the things he has taken for granted — things which might have been right for him but which might not be right for the next generation.'

A few kilometres away at Booborowie his younger brother James runs the second property. It consists of 3,900 acres of flat country in the wheat/sheep belt and it was bought in 1912 when the land round Booborowie was first put up for auction. The relatively small area of Booborowie is as important to sheep men as parts of the equally small Barossa Valley are to wine makers, although for a different reason. It is because Booborowie, although it has a rainfall of only 17 inches a year, can grow lucerne. Lucerne usually needs a fairly high rainfall to be successful and because merinos do not do well in those areas the two rarely come together. The importance of Booborowie is that here they do.

About 48 kilometres east of Booborowie is the third, and largest, of the three properties that together make up Anama. Called Caroona, it consists of 32,000 acres which were bought in 1973 when the seasons were good. With an average rainfall of only 7 inches, the country here is beyond the farming area and takes on an appearance that is all the more surprising for its suddenness. Although only half a day's drive from Adelaide, it could be almost anywhere in the Centre of Australia.

The dining room in the homestead at Anama.

Caroona consists of three different types of country. There is hilly country in the west, then patches of mallee scrub which in turn give way to the open plain country that stretches away to the east.

Whilst the properties at Clare and Booborowie are recovering from the recent drought, Caroona is not. After a run of bad seasons, the rainfall last year was a mere 4 inches and although this year will be slightly better the country still looks very dry and poor. The concern is that the bad seasons might have killed the saltbush which provides the feed in the plains country, but nobody can be quite sure until there is enough rain to prove it one way or the other. It is an anxious time for although the blackbush continues to thrive it offers far less nutrition than the saltbush.

The three properties combine to make a total of 41,300 very varied acres. They support about 15,000 sheep and each year Anama produces about 500 bales of wool and, perhaps more importantly, sells some 1,200 stud sheep.

Each of the three properties plays a different role, so that their individual features can be used to best advantage. At Clare, Ryves Hawker breeds medium- and strong-woolled merinos, whilst at Booborowie James Hawker runs the poll flock, which was started in 1947, and acts as the selling branch for the whole stud. They manage the property at Caroona together and there breed the sheep that are the cornerstone of Anama.

There are about 2,200 ewes at Caroona and they look far more athletic than those on easier country. They are large

and plain-bodied and their legs carry very little wool. They look as if they could walk for ever, and in this country they sometimes have to. They are highly developed survivors and their progeny are to be found in areas such as the Nullarbor Plain and the country round Broken Hill where sheep raising was once thought to be a quick form of financial suicide.

These ewes are mated to poll rams and the ram lambs that are born are eventually moved to Booborowie for growing out. Some of the top ewe lambs are taken to Clare as replacements for the flock there, but the rest continue to live and breed at Caroona so that their ability to survive in saltbush country is maintained. And if they are not everybody's idea of what a merino should look like, it worries the Hawkers not at all.

Ryves and James Hawker are convinced that the needs of commercial buyers are no longer served by producing huge, pampered rams in the luxury of a ram shed. Although these rams might be outstanding in their combination of traditional merino virtues, and might sell for equally outstanding prices, that is not how the buyer intends to keep them. The danger is that these outstanding rams might well turn into something a good deal more surly when they are deprived of their luxuries. When out in the paddock looking after themselves and supposedly earning their keep, they might be considerably less than outstanding.

So at Anama the rams never see the inside of a shed, instead spending all their time on open grass. Further, they will be carrying only six months growth of wool when they are put up for auction so that any defects which the Hawkers consider unimportant are clearly visible to the buyer.

It appalls the purists who think that there is no such thing as an unimportant defect and who are convinced that the beautiful appearance of a ram should overide everything else. To them, the lightly fleeced rams that have just come from the paddocks at Anama are hardly fit to be alongside the pampered specimens from elsewhere. Indeed, at one auction not long ago the Hawker brothers were told to take their rams away and to put them next door with the pigs. They did so, and sold them all.

But there is more to breeding sheep at Anama than that. Leaving them to support themselves in harsh conditions in the belief that their progeny will survive in similar conditions elsewhere might be genetically sound, but it is hardly likely to progress much beyond that. It is even less likely to reflect the changing needs of buyers. So the Hawkers work to a much more sophisticated breeding programme that is clearly defined and continually monitored.

They aim to produce heavy-cutting sheep that are not weakened by high temperatures, whose fleeces will not be subject to rot in wet weather, and which will produce a large number of lambs. This is itself a reflection of the economic needs of commercial growers, for their prosperity in turn depends on the clean weight of the fleece, its freedom from faults, and the ability of the flock to increase its numbers, or at least maintain them. The sheep must be easy care, able to combat minor attacks of internal worms and external flies, and have a strong constitution. They must also mature early and so be available for the meat trade with the Middle and Far East.

Whilst all these qualities are desirable they cannot be combined instantly in every animal. Indeed, it has long been recognised, although often overlooked, that a breeder will make more genetic progress by concentrating on a specific feature than he will by trying to do everything at once. It needs to be applied with common sense, of course. There is little point in producing the most fertile ewe in the world if her lambs have only three legs and keep falling over. Nevertheless, more progress will be made by concentrating on a small number of features than by trying to make improvements across a very broad range. They do this at Anama by sorting the sheep into family groups and then breeding within the group to develop specific features.

First, the sheep are examined visually and those with faults in the wool or serious body defects such as excessive wrinkling are rejected at this stage. The rest are then shorn and objective measurements taken of body weight, clean fleece weight, and the diameter of the fibre.

Using this information with varying degrees of emphasis, the sheep are then separated into one of three family groups. There is the strong wool family which produces rams for the more arid areas; the medium wool family for areas of higher rainfall; and the fertility family which aims to improve reproduction rate whilst maintaining an acceptable standard of wool and body traits.

The woolshed at Caroona.

A group of Anama lambs.

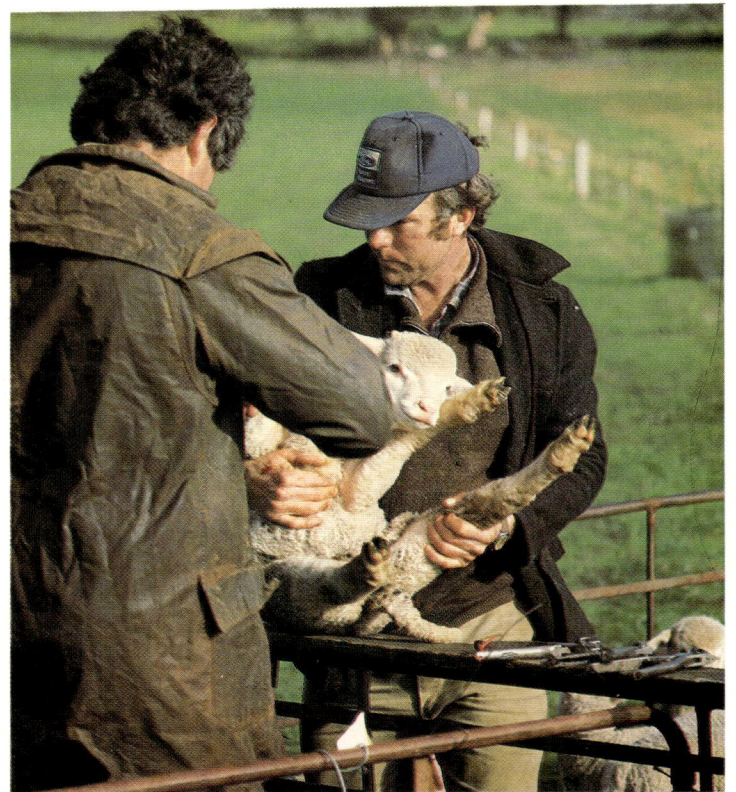

Ryves Hawker, left, working with a station hand.

James Hawker, who runs the Anama property at South Booborowie.

Ryves Hawker, who runs the original Anama property at Clare.

The deer park close to the homestead at Anama, Clare.

The woolshed at Anama, Clare.

Breeding in these groups should theoretically produce sheep that have specific advantages over those bred at random within the whole of the flock. The difficulty has always been in proving it. Breeders striving for traditional aspects of appearance have always been able to measure their progress by looking at the sheep or by sending them to compete against others at shows. But when the features that are being developed have little to do with appearance, this visual comparison offers nothing. A firm base line was needed against which subsequent improvements could be accurately measured so these could therefore be held to be a result of the breeding programme and not just a consequence of natural improvement.

The solution at Anama was to set up a separate control flock which would not be improved and which would therefore provide a standard by which the genetic progress of the stud flock could be measured. It was started in 1978 with the help of Raul Ponzoni of the South Australian Department of Agriculture and consisted of 160 ewes split into four groups of 40 ewes. These ewes were selected at random from the oldest ewes in the sud. Ewe replacements are also selected at random but revolve to the next group so as not to cause inbreeding.

Apart from mating time, the ewes and lambs of the control flock are run with the stud flock so that there is no difference in feed, climatic conditions or any variations that would make comparison dangerous. In this way, the cumulative result of breeding sheep in family groups can be measured against other sheep that have been bred at random but which have otherwise received the same treatment.

In some respects it was a dangerous step to take and certainly a brave one, for there was always the possibility that the stud flock might show little or no improvement over the control flock. Whilst in itself the lack of genetic gain might not be uncommon in merino studs, the knowledge of it certainly would have been and the consequences for Anama could have been disastrous.

Nor did anybody know exactly how much improvement would be necessary to validate the breeding programme. The improvement, if there were any, would certainly be slow for it could be advanced only slightly with each generation. Clearly the first few years would show little change, but sooner or later there would have to be a positive sign of progress. The question was how soon or how late.

In fact, it started to show sooner than the Hawker brothers expected. In the five years since the control flock was established, wool production has increased at the rate of nearly 2,000 kilograms, or five bales, every year when measured against that of the control flock. And last year the ewes in the fertility family group had an almost unbelievable lambing rate of more than 200 per cent against a still acceptable 90 per cent of the control flock and since 1978 the average difference has been about 40 per cent. Experience has shown that improvements of this kind are rarely sustained for very long, but the improvement that has been achieved then becomes the new standard on which further development can be made.

So it is not surprising that the sheep at Anama look different to the traditional merino. Indeed, Ryves and James Hawker are set on a programme that will probably change it even more in the future.

They might, in the process, also change the whole future of Australian painting.

CRANMORE PARK

It is an important day at Cranmore Park, for it is the day the shearers will come.

Nobody is quite sure when they will arrive or where they are coming from, but a few days ago the contractor sent a telegram to say that the shed they were working would cut out today and they would then travel to Cranmore Park to start shearing there in the morning.

It may seem a little casual for such an important occasion, but nobody at Cranmore thinks so. It is how it is done. They booked the contractor's shearing team some months ago and arranged that they would start this week. But shearing is not always predictable — days can be lost because of the weather and the shearing itself might take a day or two longer than expected. So when the contractor knows he is close to finishing the job he sends word to the next station to confirm their arrival. It might, as in this case, be a very brief message, but it is all that is needed.

At Cranmore Park they have been preparing for the shearing for some time but there are many things that can only be done the day before it starts. So it will be a busy day; indeed, the first of many, for there will be much to do whilst the shearing is in progress. Shearing will go on for three weeks and when it is over the shed will contain Cranmore Park's entire wool production for the year. Today the first mobs of sheep will be brought into the yards, the shearers' quarters and their kitchen will be made ready for their annual occupation, and the shearing shed will be turned into a factory for taking wool off sheep's backs.

The early morning is bright and fine. There will be no rain today and everybody is glad of that. Rain is welcome every day of the year except this one. Shearers will not work with wet sheep. If it rained today, the sheep would not dry before the men were ready to start in the morning and the day would be lost. Later, mobs will be kept in the sheds so that there will be sufficient dry sheep available even if those outside are getting wet. But there is not much they can do on the first day.

In the old timber cattle yards not far from the shearers' quarters Peter Lefroy is glad of the early sun and all that it means. But for the moment he is not thinking about the shearing, or even about sheep. Dressed in a short black jacket and with a stained felt hat on his head, he is talking to an eight-month-old foal.

Frank Wittenoom. REPRODUCED BY PERMISSION OF THE BATTYE LIBRARY, WESTERN AUSTRALIA.

He is alone in the yard, holding out some feed and encouraging her to come and take it. The foal hesitates, looking at him, then shakes her head in uncertainty. He does not move. Standing with arm outstretched, he reassures her with his voice, a gentle sound that brushes over her for minutes on end. She shakes her head again, then takes a step towards him. She stops and looks at him again and he continues his gentle encouragement. Another step forward, she pauses again, then reaches out to take the feed. He slowly moves his other hand to stroke her gently along her neck and over her head, talking to her all the time.

He is a man in his element: patient, understanding, and at ease with himself. Today, the day before the shearing starts, he knows he is indulging himself. But when you are sixty-seven and your son is running the place, you can do that occasionally.

Cranmore Park, 170 kilometres north of Perth, is one of the oldest merino studs in Western Australia. There are others that might be more fashionable, some might be more successful, and a few might think that the future is theirs alone. But none can match the achievement and innovation of Cranmore Park.

It did not come easily, for these things never do, and the soft, open country of Cranmore Park was not always so. Cranmore Park was founded by Frank Wittenoom and might, at first, have been one of the few mistakes that outstanding pioneer ever made.

He was born at Gwambygine in Western Australia in 1855. After the death of his mother, he and his elder brother Edward were sent to school in Perth. When Edward was fifteen he went as a jackeroo to his uncle Tom Burges, who had a property called The Bowes not far from Northampton.

In 1874, after spending two years working in a bank, Frank joined him there. Shortly afterwards he and Edward leased 15,000 sheep from their uncle, together with another of his properties called Yuin. Of the twenty-six men on that property, twenty were ex-convicts and, according to Frank, none the worse for that.

In 1877 Burges, anxious to take his family to England, leased The Bowes to them for seven years. During that time Frank and Edward took up a number of leasehold properties near the Murchison River, including one called Boolardy, so that when Burges returned in 1884, and the leases of The Bowes and Yuin reverted, the brothers were established landowners in their own right.

In 1891 Frank bought his brother's interest in Boolardy and, having put a manager in, he went to the goldfields at Kalgoorlie. There he became manager of one of the gold mines but in 1901 he returned to Boolardy and, with the manager due to retire, looked for a replacement.

The man he chose was his nephew, Langlois Lefroy, who had been brought up on his family's property at Walebing and who was now working as an articled clerk in a law office in Perth. Not surprisingly perhaps, Langlois readily accepted his uncle's offer and in 1903, at the age of twenty-three, he took over as manager of Boolardy. Two years later he founded a merino stud with fifty ewes and one ram bought from Boonoke in New South Wales.

In 1908, when about to leave for England, Frank Wittenoom learnt that the Midland Railway Company was planning to sell one of their leases near Walebing. It was good country, he was told, in an area of high rainfall and he thought it would be a valuable addition to the arid pastoral country at Boolardy. Thinking that it would cost about 10/- an acre, he told Langlois to buy 20,000 acres in his absence. But the land proved to be more expensive than expected and

Peter Lefroy, who pioneered many new techniques at Cranmore Park.

Bruce Lefroy, the present manager of Cranmore Park.

Arthur Clark. He walked on to the property before the First World War and asked Frank Wittenoom for a job. Now in his nineties, he still lives at Cranmore Park.

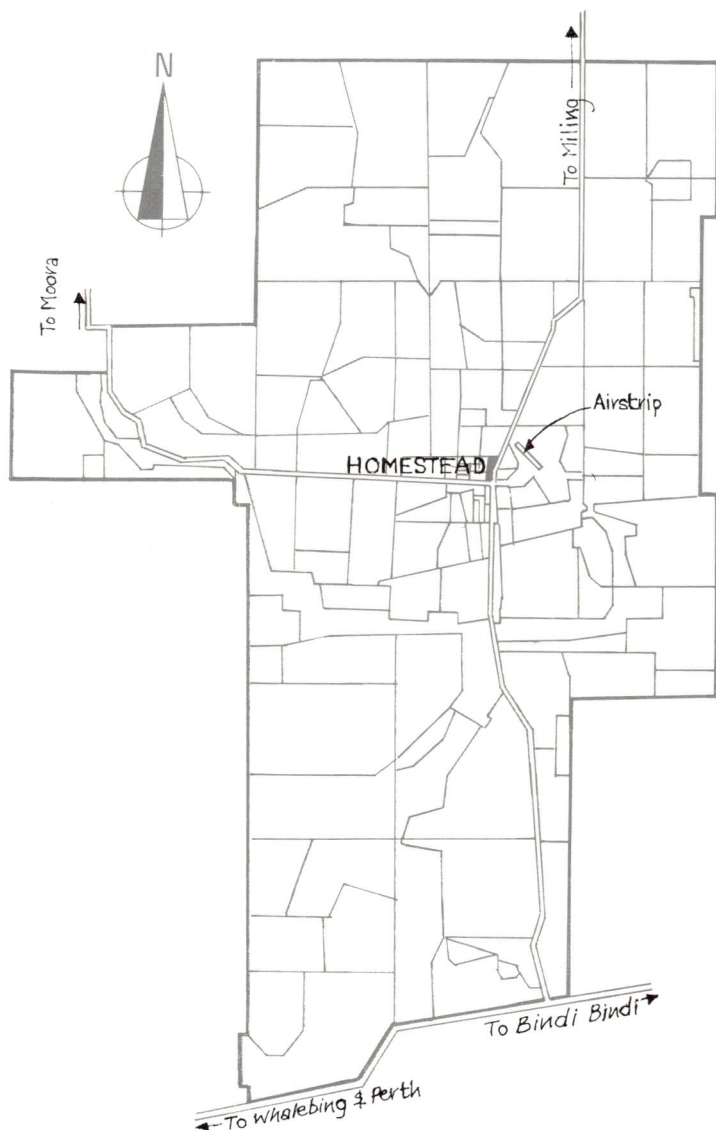

An old windmill that was used to generate electricity.

in the end Langlois bought only 10,000 acres at prices ranging from 18/- to 25/- per acre.

When Frank Wittenoom returned from England later that year he was far from impressed when he saw his new land. Heavily timbered with big salmon gums, there was hardly any natural water and little or no feed. It was, in fact, not at all suitable for what he had in mind, but having bought it he had little alternative but to use it.

He called it Cranmore Park and brought Langlois's younger brother, Ted Lefroy, down from Boolardy to establish himself on the new property and to start the awful job of clearing it. Although only a few kilometres from the family home at Walebing, Ted Lefroy lived at Cranmore Park in a tent until the first timber house was built there.

By 1910 most of the scrub and the ringbarked trees were dead and a fire was lit to burn it all off. It was not, as Frank found, a very good idea. Although the undergrowth burnt well the trees simply burnt through at ground level. By the time they fell there was nothing left on the ground that would burn the massive trunks.

The following year they burnt again and the result was not much better. Although they had ploughed fire breaks along the fences the fire got out of hand and raged towards the western boundary. Men tried to control it but soon they were forced to concentrate on their own survival. Arthur Clark, who had walked on to the property a few years earlier looking for a job (and, now in his nineties, still lives there), tells how Frank Wittenoom found himself forced back against a fence as the fire and smoke came towards him. He draped his shirt over the top wire of the fence so that his horse would see it and jumped them both to safety.

The hard work of clearing and fencing went on and by 1915 there were about 2,000 acres under crops. These had been sown as a means of improving the pasture and to provide money for further work. The stud sheep were kept there or at Boolardy depending on the season, but in 1918 the whole of the stud was moved to Cranmore Park, where it has been ever since.

In 1927, with the property now well established, Ted Lefroy took his family to Europe, which did not live up to his expectations. 'I think I know what a caged bird feels like,' he wrote. 'The monotonous roar of traffic and the purposelessness of the existence get on my nerves.'

He was happier at Cranmore Park and when he returned the following year he started to build up the stud in earnest. Already an expert sheep breeder, he was also keenly in-

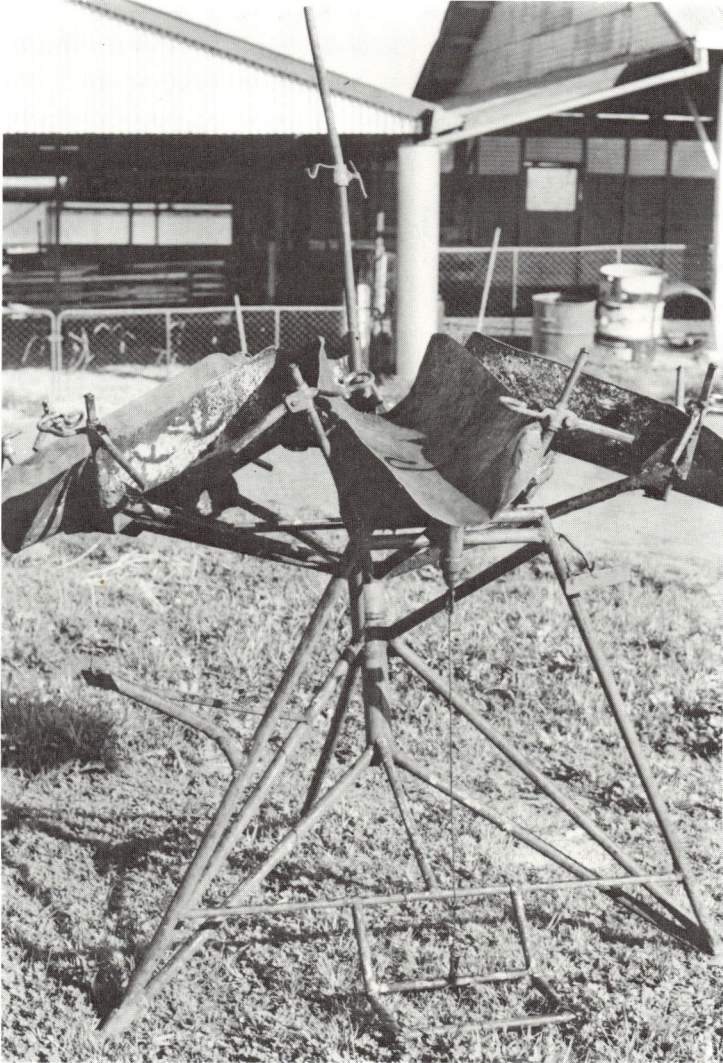

The original Cranmore Cradle. It was designed by Peter Lefroy to make the mulesing operation easier to perform.

terested in the contribution science could make and in 1933 he allowed part of the property to be used by the CSIRO for research into new methods of sheep husbandry.

Unfortunately, however, the scientists at that time had no answer to a much more urgent problem — rabbits. Frank Wittenoom had insisted on building rabbit-proof fences right from the start, although there was then hardly a rabbit within a hundred kilometres, but these had little effect when the invasion came. By the mid-Thirties the southern part of the property was virtually unuseable, although they had regained control of the northern part by digging out the warrens. 'It is a case of the rabbits or us,' he wrote. 'We must beat them or the place will be ruined.' It was to be years before they did beat them and, although the cost was enormous, they were not ruined.

Before Frank Wittenoom died in 1939 he wrote, with obvious pride, 'Assisting and watching all the development details of this property from virgin bush to its present state has given me a great interest and a great deal of pleasure and satisfaction during my late years.'

It had indeed been a remarkable achievement, but the development of Cranmore Park did not end with his death. Ted Lefroy, convinced as he was of the value of science in

sheep breeding and management, continued to develop new solutions to old problems.

He had long known, as others had, that rams selected solely for their appearance often proved disappointing as breeders. The new science of genetics also seemed to confirm that appearance by itself was probably the least reliable guide to judging a ram's potential for improving a flock.

In 1939 geneticist Dr Hagedoorn confirmed scientifically what until then had been arrived at only by practical experience. In his book *Animal Breeding* he wrote, 'The only certain way to judge an individual animal as a breeder is to judge the quality of his descendants.' It was so self-evident, he said, that it seemed absurd that animal breeders were so slow to recognise the obvious truth of it.

But Ted Lefroy recognised it and before long he had started a system of testing the progeny of rams at Cranmore Park which, with a few modifications made since, is still the basis of their breeding programme.

By the late 1940s Ted Lefroy was already using a microscope to measure the diameter of the wool fibre, having seen its importance some twenty years before it was recognised by industry as a whole. Earlier he had been an active pioneer of the newly developed technique of mulesing.

One of the problems in managing sheep in Australia is fly strike. The female blowfly is attracted to the favourable conditions provided by the stained parts of the sheep's fleece and lays a large number of eggs there. These hatch out to become a loathsome mass of maggots, all living off the sheep. They cause great suffering to the sheep and can so weaken it as to cause its death. Mulesing is an operation designed to make an attack less likely. The skin on either side of the rump is cut and it heals to leave a bare patch of skin which is no longer attractive to the blowfly.

Although the benefits of this operation were soon apparent, sheep owners were slow to adopt it because it meant that the sheep had to be handled more often and in any case it was difficult to perform on a lamb that was not co-operating. The handling was inescapable, but Ted's son, Peter Lefroy, thought something could be done to make the operation easier. So he designed a cradle like a horizontal wheel which held a number of lambs in the best position for the operation. The wheel could be rotated so that one person could lift the lambs on to the cradle whilst others did the mulesing, earmarking and other necessary work.

Having perfected it in use, Peter Lefroy refused to patent it and instead gave the design to a manufacturer in the hope that it would make the technique more widespread. He insisted on only two conditions: one was that it should be sold as cheaply as possible, and the other was that it should be known as the Cranmore Cradle. Today, mulesing is widely practised, just as Peter Lefroy hoped it would be, and the Cranmore Cradle is in daily use all over the world.

Yet another major innovation was the design and construction of the sheep yards at Cranmore Park. By the 1950s the original yards were in bad repair and Peter Lefroy decided that it would be better to replace them than to continue trying to maintain them. Timber was still scarce and expensive but he soon discovered that the war had left behind an almost unlimited supply of 44-gallon drums which could be bought for next to nothing. He devised a method of cutting and folding the metal to make rails for his new yards, then built a machine that would do it more quickly.

Next came the design of the yards themselves. From his experience in handling sheep he knew that they were more willing to walk in a curve than in a straight line. So he designed his yards so that they formed part of a big circle, with the curved sections leading to the race in the middle. Once again it was a development that was to become popular in later years and bugle yards, as they are called, are now commonplace. But when Peter Lefroy built his from 44-gallon drums they were very novel indeed.

Born in Perth in 1916, Peter Lefroy had been brought up on Cranmore Park and was educated there by Correspondence School. When he was ten he 'put on a pair of shoes for the first time' and went to school at Guildford for two years before going with his father on the trip to Europe. On his return he went back to Guildford and in 1930 he was sent to Geelong Grammar School in Victoria. Three years later, with drought and the Depression putting a heavy strain on Cranmore Park, he left of his own accord in order to save his father the cost.

After working for a stock agent in Perth he went north to work as an employee on a property belonging to his uncle Langlois and which had originally been bought by Frank Wittenoom. There he spent eleven months camped on the Murchison River trying to keep stock alive in the drought. In 1938 he went to another family property near Yalgoo and gained much experience there before his father called him back to Cranmore Park in 1944.

When Langlois Lefroy died in 1958 the Lefroys at Cranmore Park took over the running of Boolardy. With his father in poor health, much of this responsibility fell on Peter Lefroy and, realising how much he would have to travel, he decided to learn to fly.

After the death of his father in 1966 Peter Lefroy took over as Chairman of the Boolardy Pastoral Company, the company founded by Frank Wittenoom and which still owns both Boolardy and Cranmore Park. He is fit and active and very alert. Still an enthusiastic flyer, he uses a Beechcraft Bonanza to commute between the two properties and a small house in Perth.

Today, Cranmore Park consists of 12,750 acres and runs 12,000 pure Peppin merinos. They sell 1,500 stud rams each year and now, with the shearing about to start, they expect to cut about 250 bales. The property employs ten people and is now run by Peter's son, Bruce Lefroy. Forty years old, he has been managing it since 1969 when the wider responsibilities of the company meant that his father could spend less time there.

He lives with his wife and three children in the homestead that still contains the timber house that Ted Lefroy built when he came to start clearing this land. It is a constant

The circular yards at Cranmore Park. Although commonplace now, when Peter Lefroy built these from old 44-gallon drums in the 1950s they were a remarkable innovation.

reminder of those earlier members of the family who put so much of their working life into developing Cranmore Park. Bruce Lefroy has much in common with them, for it is his life too and his aims are an extension of theirs.

Ted Lefroy's aim was to produce a uniform line of sheep with the emphasis almost entirely on a constant standard of wool. There was no attempt to produce fancy rams for fancy prices whose only reliable quality might the ability to stand in a show ring. Instead, he produced commercial rams for commercial buyers who knew that those rams would breed true to type. That aim has been maintained ever since and the result is a uniformity that is remarkable in a flock of this size.

Bruce Lefroy is convinced, just as his grandfather was, that progeny testing is the only way of achieving and maintaining that uniformity and it is still the basis of the breeding programme at Cranmore Park.

There are three grades of ewes at Cranmore Park: Specials, First Studs and Generals. Each year the ram hoggetts are judged visually and the best 250 are fleece weighed and objectively measured. The results are used to select about eighty worker rams and of these, the best twelve are selected to mate with the First Stud ewes.

The lambs born to these ewes are earmarked to identify their sire and ewe and are run together until they sixteen months old, when the ram lambs are visually classed into six grades. They are marked accordingly and then drafted into sire groups so that the progeny of each parent ram can be analysed. The best four or five parent rams can then be identified and these are mated with the Special ewes. The same classing and analysing is later done with the lambs born from that mating.

The effect is to put the best rams with the best ewes and the result is a consistency that is not only visible throughout the Cranmore Park flock, but also which will be reliably carried to the flocks of commercial buyers.

This breeding programme puts many demands on the way the property is run. Because there are many different groups of sheep, for example, there has to be a large number of paddocks to accommodate them. Cranmore Park has no less than ninety-four paddocks and many kilometres of fencing. Maintaining them is a high priority for if only one ram escapes into the next paddock, that part of the breeding programme could be in ruins.

The many paddocks also mean that much attention has to be paid to water Indeed, water has always been a problem at Cranmore Park ever since Frank Wittenoom realised there was so little of it. Even digging for water was difficult. It was almost impossible to penetrate the thick granite just below the ground's surface until heavy machinery became available. Now, the property is watered by fifty dams and fifty wells.

With an average rainfall of 17 inches a year feed is usually plentiful. When a paddock needs rejuvenating, it is sown with a crop which might eventually be harvested or might be used as paddock feed. After cropping, native grasses come

A shearer bends over a sheep in the shed at Cranmore Park whilst a rouseabout carries a newly shorn fleece to the classing table.

through without further help, but clover is sown to improve the quality of this natural feed. Having escaped the drought that has gripped much of Australia, the feed is extensive and the country, gently rolling, is green and lush.

By the end of the afternoon the shearers have still not arrived, but the yards are now full of sheep, wheeling and turning as men and dogs move them from one section to the next. It looks haphazard, an uncontrolled confusion of men waving their arms, dogs racing, and sheep complaining as they are brought together, turned, then headed towards the gate into the next section of the yard. The sheep in front hesitate, stop and turn round and men call to the dogs to turn them back to the gate. The dogs scurry, the leading sheep turn and go through the gate and the rest follow in an eager stream behind.

It is not haphazard, nor is it confusing. The men make sure that each mob of sheep stays separate from the others and they move them round into vacant parts of the yards so that they can eventually lead some of the mobs into the sheds. It is like a giant game of solitaire played with living pieces.

Removing stained wool from the fleece.

When it is finished some of the sheep will be in pens in the shearing shed and others will be in pens in a shed alongside. They will be the first day's shearing and even if it rains overnight they will still be dry in the morning. The rest are in different sections of the circular yards and during the first day's shearing they will be moved into vacant pens under cover to be ready for the following day. Men will bring more sheep from the paddocks to take their place in the yards, and this will go on until the shearing is finished.

The shearers arrive after dark and hardly anybody notices the lights go on in their quarters. There are cars there now and men move between them as they make their way to the kitchen for their meal. Soon it is quiet again and the lights go out one by one until the buildings are in darkness once more. It is still early, but there will be much to do tomorrow.

It is eight o'clock in the morning and the shearing has already started. Outside the shed a few sheep are standing like peeled oranges in the small pens alongside one wall of the building. A movement, then another is pushed through a small door in the wall and it slides down a small ramp before it too stands in sudden whiteness.

Inside, it is noisy and busy. There are six shearers and each is bending over a sheep held between his legs, holding it with the left hand whilst the other runs the shears through the thick fleece. The drive shaft of the shears makes jointed angles above them until it rests on the friction wheel which gives it its motion. The friction wheels, one above each man, are on a horizontal shaft which is turned by a belt connected to a diesel engine in a room nearby.

The engine throbs steadily on and the sound of the shears adds a treble to its bass. Conversation is almost impossible in the vibrating noise but it does not matter. There is little need of it.

One of the shearers stands up and pushes the scrambling sheep through the door so that it rattles down the ramp into the pen outside. The shearer, an athletic man in his twenties, pulls the cord to disconnect the shears from the friction wheel and wipes his face on a towel hanging from a nail in the wall. The shears hang on the jointed metal shaft as he walks to the pen that holds the unshorn sheep. He lifts the nearest one under its shoulders and drags it backwards across the floor. He puts it into position, then takes the shears and pulls the cord to start them moving once more.

An old shed at Cranmore Park which was once used for butchering.

Other men move quickly around the shearers picking up the fleeces or sweeping the floor clear of wool each time a shearer goes for another sheep. At the end of the shed the fleece is thrown on to a big table and men start pulling the soiled wool from the edges. Then a tall man, the wool classer and, coincidentally, the contractor, looks at the fleece more closely, then bundles it up and carries it over to a cubicle where it joins other fleeces of the same grade.

Nearby another man is pushing armfuls of wool into a big green wool press. He stands back as the plunger comes down to force the wool into the bale. The plunger lifts and the man feeds more wool into the machine until the bale is full. He opens the machine and pulls the clean, squat bale on to a trolley and wheels it over to the scales. He weighs it, dabs paint on to it through floppy stencils, then wheels it to an empty corner of the shed and stands it upright on the floor. It is the first bale.

By the time the shearing is over it will have been joined by others, side by side and layer upon layer. And as the shearers drive off to start work on another property it will be the end of one year's cycle at Cranmore Park — and the start of the next.

THE SHEEP CLASSER

A sheep property in Australia will be one of two kinds: it will either be running a commercial flock whose wool clip provides most of the income of the property; or it will be a stud which, as well as producing wool, breeds sheep for sale to commercial growers or other studs. Some studs also run separate commercial flocks as wool producers.

A commercial flock is run for its ability to produce a profitable amount of wool and for this to continue new sheep have to be introduced to replace those that have become too old. Many of these replacements will be born in the flock but the owner will also buy in other sheep, usually rams, to introduce a new feature (such as higher fleece weight) that he might need as a result of changing economics, or simply to bring about a general improvement.

It is the function of the stud property to supply him with those sheep. In its turn, the stud is continually trying to improve its stud flock so that it can produce rams that will meet the future needs of its clients. In order to do this it has to breed rams of very high quality for use within the stud so that their features are in turn reflected in those that are sold.

Sheep, unlike many animals, pass most of their characteristics on to their children in a very pure way. If both parents carry above-average fleeces, for example, then usually their progeny will do the same. Although a lamb will combine equally the characteristics of both parents, in practice it is the ram who has most influence on the flock because he is capable of producing far more lambs than can a single ewe.

There are two ways of improving a flock. One, called compensatory breeding, balances the strengths of one parent against weaknesses of the other. The other matches like with like so that strengths of one parent are combined with similar strengths of the other. Called selective breeding, it has no room for weaknesses of any kind.

The selection of the best rams for breeding is therefore of vital importance, both in a stud and a commercial flock, and it is the job of the sheep classer to make that selection.

Dick Jago — sheep classer.

Having determined with the owner the results to be achieved, the sheep classer examines the rams whilst they are still young and selects those he thinks are most likely to produce those results. He might make this selection entirely by eye or he might use objective measurements such as fibre diameter or fleece weight with varying degrees of emphasis.

His judgement has to be particularly sound with a stud flock. A stud relies heavily on its reputation, often built

A magnificent South Australian stud ram.

up over many years, as a reliable supplier of breeding stock that can be expected to introduce their desirable qualities to their clients' flocks. If the classer makes a mistake, or anticipates a demand that does not eventuate, it might be years before the resulting damage can be repaired and the stud might have little reputation left.

Dick Jago is one of the best sheep classers in Australia. Born in 1933 at Forbes in New South Wales, he joined Haddon Rig in 1953 as a jackeroo. Malcolm McLeod,

who was then classing the Haddon Rig stud, saw his potential and gave him special training. When McLeod retired in 1971, Dick Jago took over the classing at Haddon Rig and has been doing it ever since.

A firm advocate of selective breeding, Dick Jago is, like most classers, a self-employed freelance. Based in Dubbo in New South Wales, he now classes sheep in most States of Australia and examines some 350,000 stud and flock sheep every year.

TERRICK TERRICK

Although much of Queensland is given over to cattle, part of it remains firmly committed to sheep, continuing a tradition that has lasted generations. Centred around the towns of Longreach and Blackall in the west of the State, it lies just south of the Tropic of Carpricorn and is not far from the northern limit of sheep-raising in Australia. It is harsh country which has demanded much from those who use it.

Terrick Terrick, one of the most famous merino studs in Queensland, is 67 kilometres from Blackall along a road that is mostly open black soil and which rapidly becomes impassable in heavy rain. It is a lonely, empty road through scrub and occasional paddocks, with only a few gates to indicate the presence of other properties. Alongside it, but not always visible, runs a single rail track which ends as a siding about 25 kilometres from the entrance to Terrick Terrick.

There, standing incongruously in the middle of nowhere, are two rail tankers. Old and scarred, they look as if they have been there for years, forgotten by the managers in far-away Brisbane. But they have not been forgotten. Indeed, they have been there for only a few days. Their presence reflects the vagaries of the climate in this part of Queensland. They contain molasses. Thick, sweet-smelling and sticky, it will be mixed with a protein supplement and

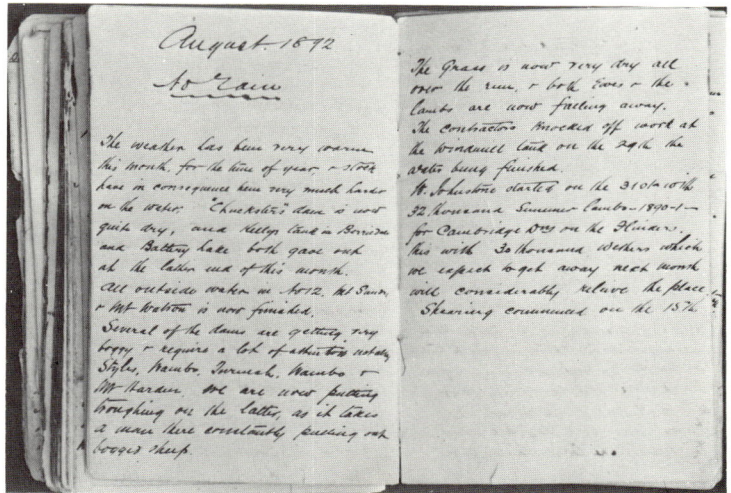

The diary kept by Jonathan Dickens at Terrick Terrick between 1882 and 1898.

A team of shearers doing their washing at Terrick Terrick in 1924.
REPRODUCED BY PERMISSION OF THE JOHN OXLEY LIBRARY, QUEENSLAND.

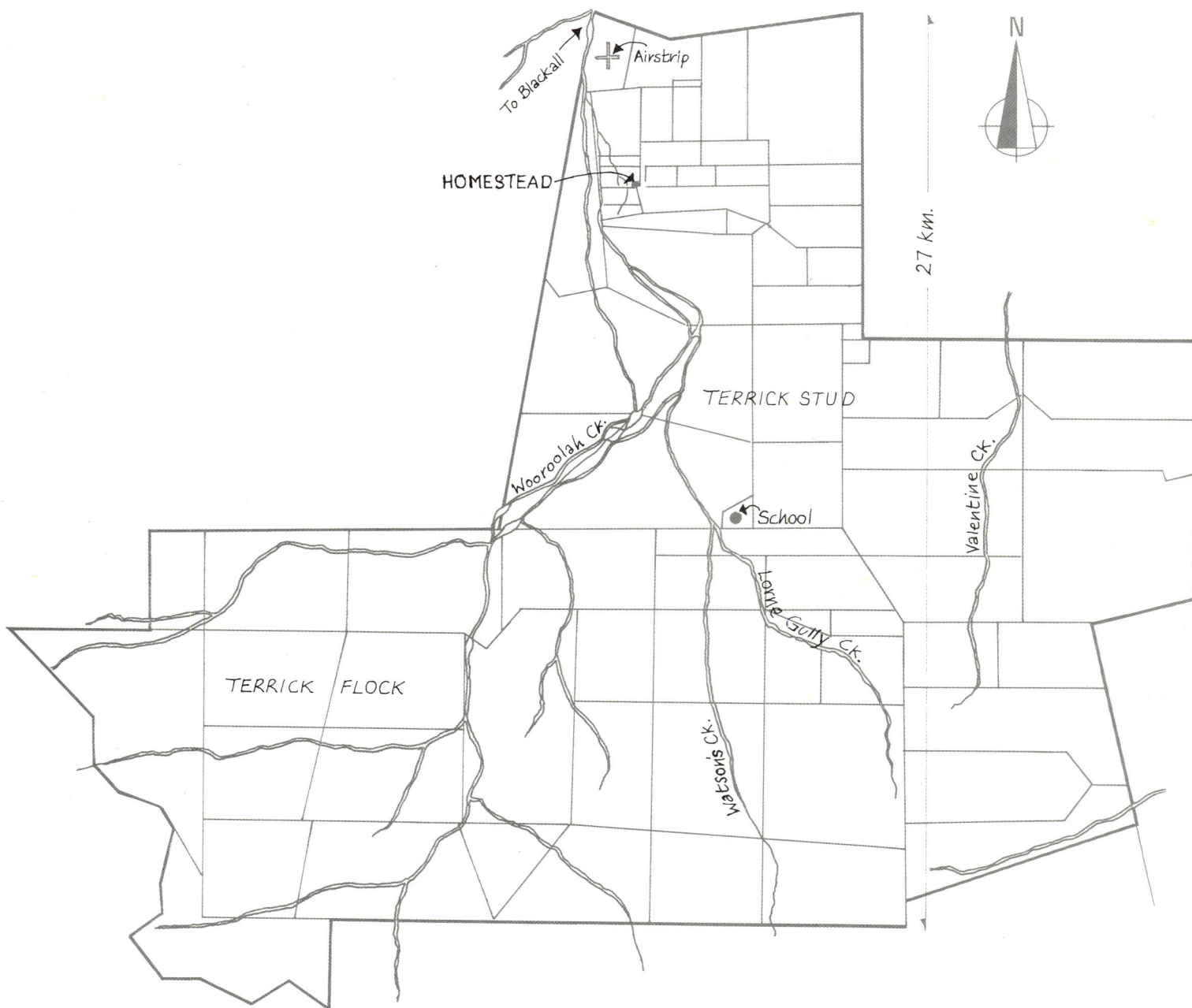

To Blackall

Airstrip

HOMESTEAD

N

27 km.

TERRICK STUD

Wooroolah Ck.

Valentine Ck.

School

Larrie Gully Ck.

TERRICK FLOCK

Watsons Ck.

put out for the sheep in the paddocks of Terrick Terrick. In this drought-stricken summer the sheep are unable to find enough growth in the paddocks and have to be fed by hand. It is not the first time they have had to do this at Terrick.

This part of Queensland was slow to be settled and it was not until 1864 that John Govett sought permission to occupy this land. He ran it with his partner James Thompson until 1874, when it was taken over by Thomas Russell. During the next four years Russell acquired leases for a number of adjacent runs and in 1880 these were transferred to Charles Rome. By the following year Rome was the sole lessee of eleven runs and he consolidated them into one run that became known as Terrick Terrick.

Those early years were marked by almost continuous drought and the need to preserve water was soon obvious. But the conditions of the lease did little to encourage the large investment that this required. Rome, like most land-

owners in the area, stumbled along as best he could. But not for long, for soon the property changed hands again.

Richard Gardiner Casey already had some pastoral experience, having managed a sheep property until it was sold in 1883. Casting round for his next job, he was flattered when Donald Wallace invited him to go into partnership with him. Wallace was a prominent Victorian and in 1882 had raised finance in England to acquire four properties, of which Terrick was one. Casey would run Terrick and from there he would also supervise the other three.

Casey was thirty-seven when he moved to Terrick in 1883. At that time it consisted of 535,000 acres and carried about 140,000 sheep, as well as a few thousand head of cattle and two hundred horses. The homestead was built of heavy logs and was almost like a fortress, with small loopholes in the walls, for protection against attacking Aborigines.

By the following year Casey could already see that his job would not be an easy one. There had been very little rain

Rob Beatty, manager of Terrick Terrick.

Cathy Beatty.

Lizzie Beatty doing a lesson with the School of the Air.

Graham Eagle. He has been breaking horses at Terrick Terrick for thirty years.

and, with nothing to support them, nearly a third of the sheep had died. It was also clear that the interest on the money Wallace had raised in London was more than the property could hope to pay and it was decided that Casey would go to London to try to renegotiate the loan. He was away for the whole of the second half of 1884, but his trip was unsuccessful. The lenders refused to alter the terms and Casey returned convinced that they were simply waiting for them to fail so that they could take over the property.

Against the ever-present threat of foreclosure, Casey got on with the job of running Terrick Terrick. In 1885 he bought over 1,500 rams from Boonoke and transported them over 3,000 kilometres to Terrick. It was a brave attempt to increase the quality of the Terrick flock (it was not yet a stud), but the following year the rain failed again. He had twenty-five men cutting scrub to try to keep the sheep alive. 'When,' he wrote, 'are our troubles to end?'

They ended, in part at least, sooner than he expected, for with the unpredictability of the Queensland climate the following year was exceptionally good. Grass was stirrup high and the number of sheep increased rapidly. In October they shore 188,186 sheep for 3265 bales and although wool prices were low the year ended on an optimistic note.

It had been an exceptional season, and it was also short-lived. By the middle of 1888 Terrick Terrick was dry again and unable to support the increased stock. The rate of this increase had been remarkable. Although he now sent nearly 50,000 sheep out on to the roads in search of feed there were still nearly 200,000 left to pick a tenuous existence from the empty paddocks.

With no buyers, Casey found it impossible to reduce the stock and was well aware that he was running far beyond the carrying capacity of the land. Indeed, he had little choice, for with the debt mounting up in London he had to gamble on getting better seasons. In 1891 they shore 320,000 sheep at Terrick but with wool prices falling there was no way to make ends meet. The following year he was sending sheep to Barcaldine to be boiled down for tallow.

In 1893 the partnership was dissolved and Casey's nightmare was over. During the ten years, sheep numbers at Terrick had ranged from 120,000 to 320,000 and the debts had increased from £350,000 to £650,000. Casey left Terrick with little more than the clothes he stood up in, but he was glad to leave all the same. It had been a disastrous experience.

Terrick Terrick was now taken over by the Australian Estates and Mortgage Company and in 1896 they formed the Terrick stud by selecting 850 ewes from the flock. There were at that time nearly 100,000 ewes in the flock and the selection is an indication that the quality was not very high.

At the same time a number of stud rams were bought from Boonoke and joined with the selected ewes. This programme continued until 1916, after which pure Peppin rams were bought each year from a number of studs in New South Wales. Then in 1929 the company decided that future rams would come from its own studs of Raby and Oolanbean No

A windmill seen through the door of the woolshed.

1. Meanwhile, pure Peppin breeding ewes were introduced from Wanganella, Munnell, Raby and Oolanbean.

The rams born at Terrick were used in the stud and were rarely sold to other breeders. That changed in 1931 when the company negotiated a new stud lease with the Queensland government. Under the terms of the lease the stud area was increased and could now be run separately, but one of the conditions was that rams had to be made available to outside buyers.

In addition to developing the stud, the company spent thousands of pounds on improvements to the property. In Casey's time water had been stored in dams built across the creeks to provide more or less permanent waterholes. Known as Chinamen's dams, they had been built by teams of Chinese who had left the worked-out goldfields of Queensland. Many of these were now repaired and others built. Artesian bores and sub-bores were also sunk and the water led through the paddocks in a series of open bore drains.

In 1941 a poll stud was started at Terrick with a hundred merino ewes and two sports which had turned up in the merino stud and by 1955 this stud had grown to 300 poll ewes. At that time an outlying part of the property called Gowan was running the Terrick single stud ewes and when a

The station cook at Terrick Terrick.

Ivan Oliver, overseer.

shearing strike in 1956 made it impossible to shear them there they were walked to the Terrick shed for shearing. Whilst they were there 2,500 merino ewes were selected for poll breeding and it was these sheep, together with the Terrick poll sires, that were later walked back to Gowan. They were the start of the Gowan poll merino stud.

Australian Estates was a big company with interests in many fields and they continued to run Terrick Terrick very successfully until 1974, when the company was acquired by another huge organisation, CSR. Although perhaps better known as the leader of the Australian sugar industry, CSR had long been actively involved in the pastoral industry as well. So whilst the acquisition of Terrick Terrick was not the main reason for CSR acquiring Australian Estates, there was never any doubt that, having taken it over, they would continue to develop it as a merino stud.

Today, Terrick Terrick consists of 160,000 acres and carries about 40,000 sheep. It consists of two properties, Terrick Stud and Terrick Flock, which are now run as one. Gowan is run as a separate CSR property and continues as a very successful poll stud.

A few kilometres from the railway siding is the entrance to Terrick Terrick, with a modest sign alongside the gate.

There is no passing traffic in this lonely spot and the sign is merely a reassurance to visitors that they are not as lost as they thought.

The road curves over a ridge and from the top one can see the buildings of Terrick. On the right is the woolshed and the shearers' quarters, and about a kilometre ahead are the neatly grouped buildings of the homestead area.

It is a small village grouped round an open space. On the left is the homestead. Built in the 1950s, it carries the unmistakable look of Queensland, with a big covered verandah protected on the outside by metal slats that make a permanent Venetian blind. Beyond the homestead are the jackeroos' quarters, the mess and the cook's house, whilst the rest of the square is made up of more houses, the maintenance sheds and the overseers house. In the middle of the square is the station office, but it is hardly ever used. There is a smaller office in the homestead and Rob Beatty, the manager of Terrick, prefers to use that when he can postpone the paperwork no longer.

Born in 1947, Rob Beatty went to school at Nudgee College in Brisbane before becoming a jackeroo at the Clark and Tait property at Mantuan Downs in 1966. From there he went to Barcaldine Downs and Bimerah before becoming

The jackeroos' quarters.

overseer at Norwood near Blackall. The following year he was appointed manager of Mount Harden and then in 1974 he went as overseer to Goolgumbla, a merino stud near Jerilderie in New South Wales. When this was sold in 1979 he became manager of Gordon Downs near Capella in Queensland and in 1981 he joined CSR as manager of Terrick Terrick.

He is a big man with an easy-going nature which tends to disguise the fact that he is a highly experienced and professional manager. He sees little point in exerting authority when it is not necessary, but when it is there is no doubt that he is the boss.

Since taking over at Terrick he has embarked on an extensive programme of improvements, many of which were long overdue. In twelve months he has spent about $350,000 on improving water storage alone. Turkey's nests have been built to provide a more reliable supply of water to the troughs than was possible with open dams. The old metal troughs themselves are being replaced with concrete ones that look as if they will last for ever.

The sheep have also received considerable attention. When Rob Beatty took over, he and Duncan McDonald, who classes sheep at all CSR properties, ruthlessly culled those that were not up to standard. In 1976 a ram was bought from Collinsville in South Australia and others have recently been bought from Goolgumbla. The aim is to produce big, heavy cutting sheep with a good constitution suitable for the conditions found in Queensland and parts of New South Wales.

Unlike many merino studs, they remain firmly committed to showing sheep and in 1982 they won the prize for the supreme Queensland exhibit at the State Sheep Show at Blackall. They show regularly at the most important shows in western Queensland and see it as the best way of keeping their name in front of buyers. It is an expensive business but

Dawn muster.

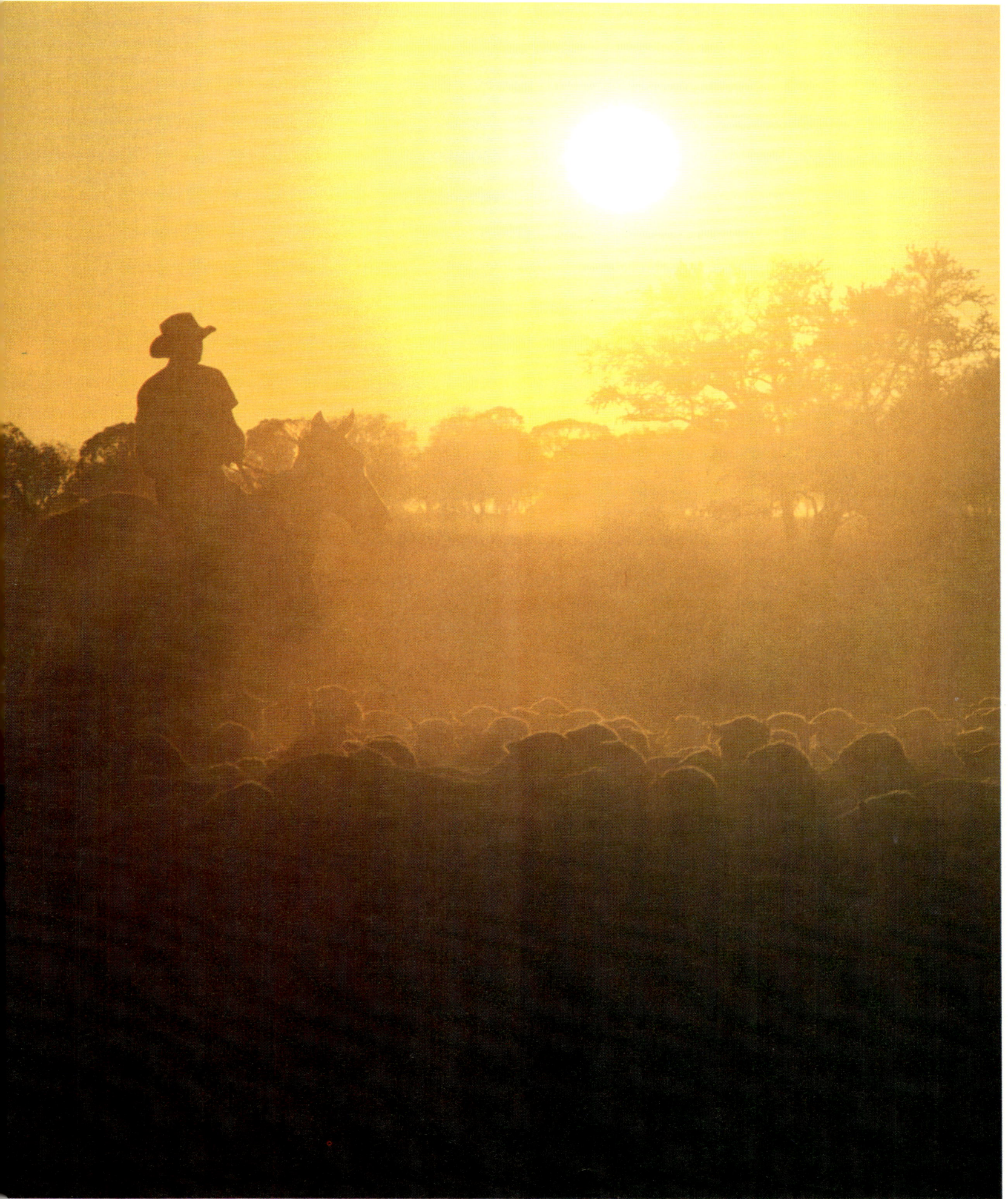

they think it is worth it — and they sell about 1,100 rams a year from Terrick and about 1,700 polls from Gowan.

There is a staff of twelve at Terrick, including four jackeroos. With families, about thirty people live on the property and Rob Beatty is sometimes as much a mayor as he is a manager.

Single people eat in the mess, whilst families look after themselves using supplies provided by the company as part of their wages. Stores come out from Blackall once a week and are issued every Thursday. Milk comes from several dairy cows and six sheep are killed every week to provide an abundance of meat. Mail comes and goes once a week and with the mail come the newspapers — a week's supply on one day. But they are important to everybody on the property for no Queenslander can live happily without the *Sunday Mail*, even if it doesn't arrive until Tuesday.

The nearest doctor is 67 kilometres away at Blackall. Fortunately accidents and serious illness are rare, but they can happen at any time. The worst time would be when the road was impassable and the phone down. They could use the radio to call an air ambulance but if conditions were that bad it might not be able to land, in which case they would been given medical instructions over the radio.

Cathy, Rob's wife, is bookkeeper, store keeper, and sometimes wishes she had nursing experience as well. Children are the main anxiety. They live a life that would be the envy of their urban counterparts but it is not without its dangers. It is not so much a matter of keeping danger from them as keeping them away from it. A young child sees no danger in being in a yard with a tail-swishing stallion and walking up behind it with a stick.

When that happens the behaviour of an adult is almost as critical as that of the child, for any attempt to dash recklessly into the yard and rescue the child is likely to be disastrous. Instead, the child would be told to put the stick on the ground and to walk slowly backwards to the fence. Only then can the child be lifted out of the yard from a danger that the child never suspected but which was only too obvious to everybody else.

Despite the passing of time, Casey would still find much that was familiar. For example, Rob Beatty has little time for 'gasoline cowboys' on motorbikes and instead prefers to work the stock with horses. They are quieter, so you can hear animal noises when mustering, and in good seasons when the grass is 'stirrup high' you can see more from a horse than you can on a bike.

There are about a hundred horses on Terrick and every year Graham Eagle arrives to break in the three-year-olds. He has been breaking Terrick horses for thirty years and can probably remember most of them. He has just saddled a filly for the first time and he is slow and gentle as the filly stands on tip-toe, ready to rebel.

'Now be a nice girl,' he says. 'Be a nice girl, like your mother. Be a nice girl...' He keeps repeating the soothing sound as he swings into the saddle and by the time the filly

A Terrick Terrick ram.

realises what has happened it is too late. She bucks, rears and jumps but it is no good — very few horses beat Graham Eagle from this point. When he dismounts he is still talking to her.

Casey would also recognise that drought once again afflicts Terrick. Now, nearly a hundred years later, Rob Beatty is facing almost exactly the same problems that tested Casey so severely. He has untold advantages, to be sure, but he is still faced with a reality as harsh now as it was then: how to keep sheep alive when there is not enough for them to eat.

One answer is hand feeding, and that is why the tankers are standing there on the rail siding. Unloading the molasses is long, dreary work. Then it has to be taken to another tank on the property and unloaded again before it can be taken in smaller quantities to the feed bins in the paddock.

It is 4.30 in the morning and even though it is still dark the temperature is 25 degrees. In the mess Ivan Oliver, the young overseer at Terrick, is cooking a couple of pieces of steak and some sliced tomato in an electric frying pan. Before it is ready he is joined by Peter, a fifteen-year-old jackeroo who has been on Terrick only a few weeks and who has just discovered for the first time that there is a 4.30 in the morning.

They sit at a long table and eat in silence. It is quiet outside but there are lights on in some of the houses. They finish breakfast and Ivan gets up to start the day. As he walks out of the mess and jams his hat on, Peter realises that he should be doing the same. They are going out to move sheep from a bare paddock to one that still has a little feed.

By the time Peter catches up, Ivan is already at the stables and saddling his horse. Peter does the same and from the darkness comes the familiar sound of horses whinnying and the soft chink of harness. Ten minutes later the horses have been loaded into the high-sided trailer and Ivan starts the engine of the Toyota. He resists the temptation to gun the engine and wake everybody up and instead pulls quietly away, the headlights sweeping round as he turns on to the track to the paddock.

The Bore Cafe. Ivan Oliver fills the billy with boiling water pumped from under the ground by a nearby artesian bore.

It is 25 kilometres away and by the time they reach it the dawn is glowing in the sky. They unload the horses and Ivan tells Peter to ride down the nearest fence line and drive the sheep to a turkey's nest about 2 kilometres away. Ivan will muster the rest of the paddock and join him there. They ride off in their different directions, leaving the Toyota and trailer straddling the track.

An hour later it is daylight and Peter is at the turkey's nest with a mob of a hundred sheep. He is sitting on the ground near his horse as the sheep wander around in a loose mob. Occasionally they break into smaller groups and start to wander off and Peter remounts and drives them back. There is no sign of Ivan and Peter keeps looking across the paddock for him. The far fence is about 3 kilometres away and out of sight because of the scrub and the rising ground. It is now nearly eight o'clock and he is alone and far from anywhere with a mob of sheep and his horse. But he has been told to wait, and that is what he does.

Half an hour later he hears the bleat of sheep in the distance and soon he can see them. They are walking in a long line abreast over a distant ridge and as they come nearer he can see Ivan riding easily behind them. There are more than 500 sheep in the mob and they are doing exactly as Ivan wants. At this moment, far away, a city worker is probably having considerably more trouble walking one dog.

Peter checks that his sheep are close together, then rides off to join Ivan, skirting wide round the mob to avoid breaking them. They ride easily together, not talking much as they drive the mob to join the smaller one by the turkey's nest. Peter dismounts and opens the gate into the next paddock and they both count the sheep as they go through. When they have finished they have a difference of one in the count and Ivan writes the number in his notebook. It is Peter's count that he records because even though he is only in his twenties he already knows a lot about men.

They close the gate and ride back to the Toyota. It takes them only a few minutes to load the horses on to the trailer and soon they are heading off to another paddock about 8 kilometres away.

Splitting up as before, they soon have the sheep collected in one corner of this fairly small paddock. They drive them through the gate into the next paddock and count off a few hundred that will remain there. They then drive the rest back into the first paddock and count them through the gate. This time they have identical counts and Ivan again writes the number in his book.

By the time they have finished it is 10.30 — smoko time. They load the horses again and drive off, Ivan scanning the country as they go, evaluating feed and the condition of the sheep. Fifteen minutes later he swirls to a stop and gets out. Peter does the same, hoping that they will be having a break but not sure why they have come here. Ivan swings a blackened billy from the back of the Toyota and shows him why.

They are at the head of an artesian bore and a few metres away a pipe is gushing boiling water into a small, bubbling pool at the top of the bore drain. Ivan holds the billy under the pipe and fills it. Then he throws in a handful of tea, hits the side of the billy with a stick, and pours some tea into Peter's tin cup which he has taken from his saddle. They call it the Bore Café, and it serves good tea.

By the time they get back to the homestead it is 11.30 and very hot. They have been working hard for nearly seven hours. After seeing to his horse Peter can relax and wait for dinner, but he will be needed again at 2 o'clock. It is the end of the morning, not the end of the day.

Whilst Ivan and Peter were in the paddocks, Lorna Armstrong has been teaching school. She has four pupils: three Beatty girls and the son of one of the station hands.

At eight o'clock, neatly dressed in school uniform, they pile into her car outside the homestead and she drives the 15 kilometres to the station school. It is a small building standing next to an empty cottage and it is surrounded by dry paddocks that disappear into the distant trees and scrub.

Inside is a single room, but it is unmistakably a classroom. It has a blackboard, desks, books, and children's paintings on the walls. Most teachers would find it totally familiar, even though they might be disturbed by the apparent wilderness outside.

Lorna lets the children burn off a little energy outside before calling them in to work. They sit down and she makes sure each child knows what to do. They are working through a course set by the Primary Correspondence School in Brisbane and which comes out in the mail. It is skilfully designed to provide primary education in remote areas such as this and allows a mother to teach her children if there is nobody else to do it. Each week the child completes a paper and sends it back to Brisbane for marking. It is a simple system, but its simplicity disguises the considerable skill and dedication that goes into preparing the papers.

After a mid-morning break Lorna switches on a communication radio in one corner of the room and young Lizzie sits eagerly in front of it, holding the small microphone. It is time for School of the Air. Run from Charleville about 300 kilometres away, it provides each age group with half an hour's school each day. The teacher opens the class and calls up each child in turn.

'Good morning, Lizzie.'

Lizzie pushes the button on her microphone.

'Good morning Miss Randall, girls and boys.'

Most children answer, some do not. Some of those who do can be heard clearly, others are drowned out in static. They might be on the other side of Charleville on properties 700 kilometres from Terrick.

Having called the role, Miss Randall asks everybody to open the story book at page 10.

'James, will you please read the first two paragraphs for me.'

Static surges in and out as a small piping voice reads aloud.

'That was well read, James. Lizzie, would you please read the next two paragraphs.'

Lizzie pushes the button again and reads into the microphone, stumbling over some of the words but getting them right in the end.

'Very good Lizzie. Now, Michael...'

After everybody has read aloud Miss Randall says she will read them a story and then ask questions about it. Lizzie listens and then waits for the first question.

'Why did the dog escape?'

Lizzie hits the button and calls 'Lizzie!'

'Yes Lizzie.'

'Because somebody left the gate open, Miss Randall.'

'That's right, Lizzie. Now, who...'

The thirty minutes is soon over, but the teacher has covered a great deal of ground in that time. As they all sign off Lizzie puts down her microphone. She still has a far away look in her eyes for she has been a long way from Terrick and she is not quite back yet.

Outside, the sun is gearing up for the full heat of the day. The trees hang limp and sheep are camped beneath them in the shade. In the distance a ute leaves a trail of dust and elsewhere men are filling bins with feed that will keep sheep alive.

In the classroom Maree Beatty leaves her desk and takes her exercise book to Lorna Armstrong.

'Miss Armstrong, I've finished that essay.'

She goes back to her desk. Maree is ten years old and she has been working on this essay for several days. Lorna suggested that she might write about something she was familiar with, instead of trying to imagine a world she had never seen. With the class busy at their desks, Lorna opens the exercise book.

Lorna Armstrong with her pupils outside the schoolroom at Terrick Terrick, some 15 km from the homestead.

A SEVERE DROUGHT AT TERRICK

'As what is our usual wet season comes to a close, we are experiencing one of the worst droughts in this century.

'Sometimes we drive out with Daddy and all we can see are big whirlywinds and dust storms over the Downs, which have, in the last twelve months, turned into sandy and stoney plains.

'Meanwhile, men are kept busy feeding the sheep that are nuzzling in the dirt looking for a tiny blade of grass and digging up seeds and roots which they will gladly eat.

'The sheep look like walking skeletons with wool. There are bones and carcasses everywhere as many sheep have died of starvation.

'Now many waterholes have dried up and are very boggy, therefore we must, instead of carting water to the stock, move them into a different paddock where the waterhole is not boggy.

'This drought has made a change to the wildlife on Terrick as many more birds have come into our garden to drink the water and eat the grubs from our lawn. Many kangaroos and emus have died of thirst and hunger. Our house dam has also many birds but they are mostly ducks and brolgas.

'Just here on Terrick Stud we have not yet had to let our lawns die. But at Terrick Flock they have had to because there is not enough precious water.

'My family like every other family on the land is praying to receive some of the relief rain that we hope will be coming soon.'

EAST BUNGAREE

It is Tuesday morning and it is raining at East Bungaree. Yesterday was warm and sunny and the breeze pushed fluffy white clouds across the blue sky so that their shadows covered the ground with a moving pattern of light and shade. Now, there is no sun, no light and shade, and yesterday's summer brilliance has reverted to winter monotony.

The gentle rain, hardly more than a drizzle, started so softly that at first it was barely noticeable. The heavily overcast sky hung moisture in the still air to create an unpleasant dampness that was still not rain. Then the wind came, a cold north-easter that pushed darker clouds below the grey ceiling and turned the dampness into definable rain, slight but persistent. The ground turned darker and the trees, wet and glistening, sighed in the wind and rustled together to give the impression of much heavier, more welcome rain.

'Will this rain do any good?'

'No — it will hardly register.'

At Old Belcunda the rain washes the walls of the old shearing shed and in the yards little beads of water glide slowly along the edges of the rails and drop off one by one. Dogs bark and jump on to the backs of soggy sheep to move them forward until the race is full of a single file of miserable ewes and lambs. The men, startlingly colourful in glistening oilskins, close the gate behind them and move along the race. Using a hand gun, they squirt a measured amount of liquid into the mouth of each animal as a protection against worms and other internal parasites.

When all the sheep have been treated Brian Jamieson, the overseer of East Bungaree, goes to the head of the race and unfastens the gate. The sheep shuffle towards him and he swings the gate from side to side to direct each animal into one of the two pens. Ewes and young lambs go into one, the older lambs into the other. Now big enough to look after themselves, they will be put on to good feed in another part of the property whilst the ewes and younger lambs will be returned to their paddock at Old Belcunda. The ewes,

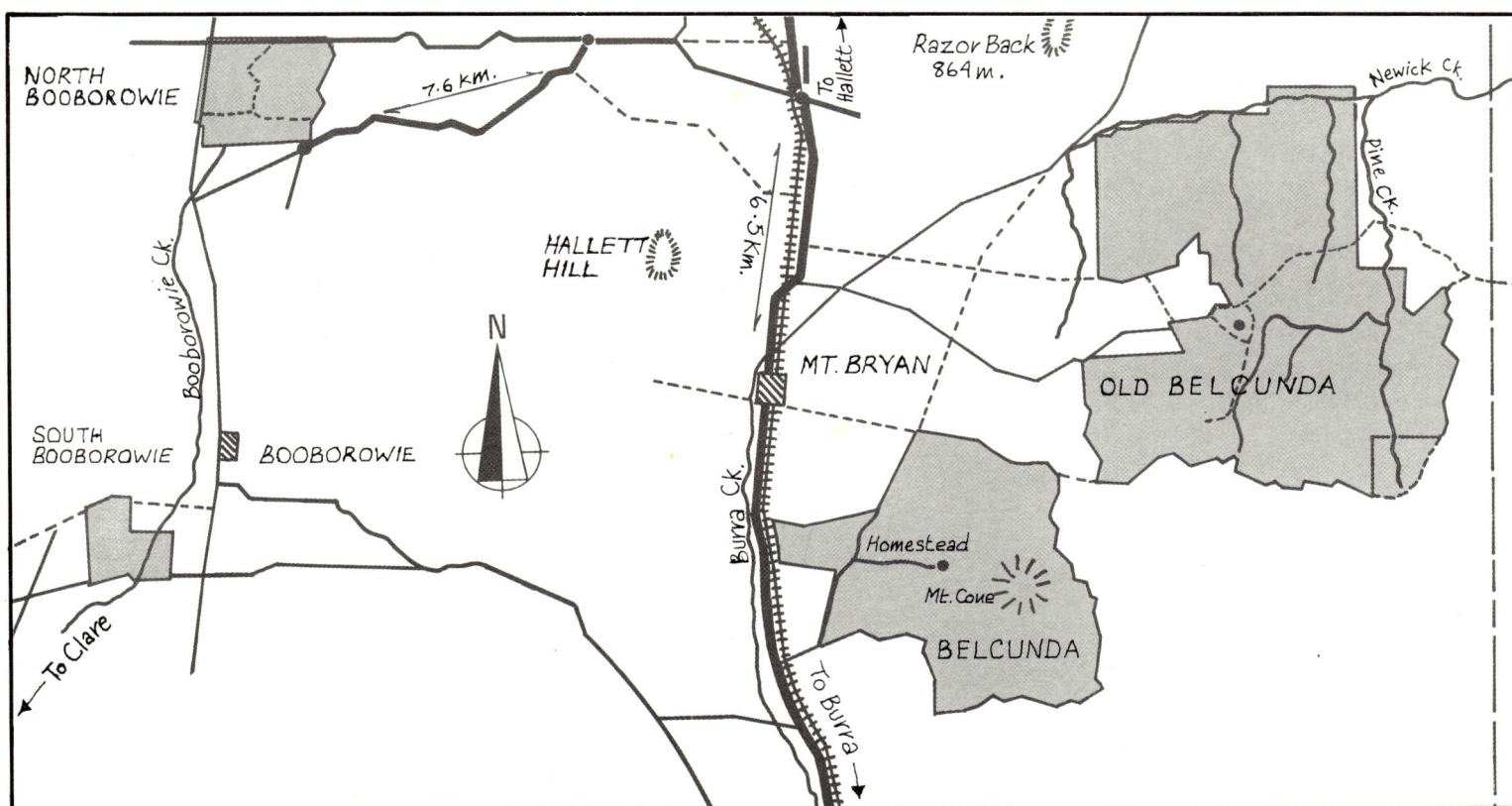

relieved of the older lambs, will be able to build themselves up and take better care of the younger lambs which still need their support.

They have had a very good lambing at East Bungaree and because it was in two drops several weeks apart some lambs are more advanced than others. But this increased population has put too big a strain on the pasture and it can no longer support them all. For, unlikely though it seems on this soggy day, East Bungaree is in one of the worst droughts it has ever known.

A drought is not, as city people often believe, a period of no rain: it is a period when the rainfall is so far below average that there is no longer any growth. The problem is not that sheep die of thirst, for most properties have bores that draw water from the almost unending supply below the ground, and dams which trap the small amount of rain that does fall. The problem in a drought is lack of feed.

In the first year there will usually be enough feed in the paddocks to keep the stock alive. It will be old and dry and look unpalatable, but it will still provide nutrition. But in the second year the problem becomes much more serious. By then most of this feed will have been eaten and, because of the low rainfall, there will be no new feed to replace it. The paddocks become bare and the dry soil, no longer held by a mass of roots, is swirled away in every breeze, taking with it the very fertility that will be needed when the rain at last returns.

Torrential rain, that might be thought to be breaking the drought, can do even more damage. It beats the surface of the ground into an unyielding, solid mass and carries away even more of the already diminished top soil. This devastation can be so severe as to last for years, so that men on the land might still be coping with the effects of the drought long after city people have forgotten all about it.

East Bungaree is now in the second year of drought. The paddocks which would normally have feed a metre high are now empty and scraped clean by the wind. Moving sheep to other parts of the property that are not so badly affected gives those remaining a better chance, but even then they cannot survive on what they can find. Feed has to be taken to the paddocks to keep them alive, emergency rations that have already cost $200,000 and which somehow will have to be maintained until this drought is over.

This drizzling rain will not break it. The rain will wash the dust from the trees and for a while the loose soil in the paddocks will not be so vulnerable to the wind. If it goes on for a day or so it might even produce a little growth, a mere tinge of green between the now useless stumps of earlier grass long since eaten. The fragile new leaves, called a 'pick' by sheep men, are always welcome but they do little to fill the bellies of hungry sheep.

At least two inches of rain are needed to break this drought and it would have to fall gently over a period of days so that it can soak into the ground instead of cascading away in rushing streams. And there would need to be as much again about a month later so that the feed that has

A team of East Bungaree rams ready for the 1968 Perth sales.

germinated can grow to maturity and once more cover East Bungaree with the rolling greenness it so desperately needs.

East Bungaree was founded in 1906 when the huge property of Bungaree, founded in 1841 by George Hawker, was divided amongst his sons. Concerned by a threatened tax on large landholders, and in any case keen to go their separate ways, the brothers agreed to subdivide Bungaree's million acres into equal parts and to distribute them amongst themselves by drawing lots.

It was E. W. Hawker who drew the eastern part of the property. He called it East Bungaree and stocked it with his share of the famous Bungaree stud which Hawker had originally founded on Rambouillet rams imported from France. In 1918 he sold most of the land for use as soldier settlements and bought land nearby which he considered more suitable for raising the big South Australian merinos which he had now developed. In 1921 his son Walter took over the running of East Bungaree and he continued to breed large sheep with plain bodies and a fleece of strong wool with a long staple. At an Adelaide auction in 1925 East Bungaree set a record price for greasy wool that stood unchallenged for the next twenty years.

In 1954 Charles O'Connor joined East Bungaree as overseer and later took over from Bruce Crockett as manager. O'Connor soon made his mark as an outstanding sheep man and under his energetic and enthusiastic guidance the stud was soon recognised as one of the best in the country.

But when Stanley Hawker died in 1979 it looked as it the years of development and achievement at East Bungaree

about 15 kilometres north of the small town of Burra. A few kilometres further north is the biggest of the properties, Old Belcunda, which consists of 10,842 acres of grazing hills. Rising to nearly a thousand metres, this high, sweeping country is normally rich in feed but is now the most seriously affected by the drought.

To the west, in the rich lucerne-growing country of nearby Booborowie, are the two other parts of East Bungaree. There are 576 acres south of the town, whilst further north is another property of 1,302 acres which is the part of East Bungaree that most of its clients see. It is rich and fertile country and its greenness, even now, is a welcome contrast to the starkness of the Old Belcunda hills.

The four areas of East Bungaree together make a total of 18,473 acres and they carry about 10,500 South Australian merinos. In good years the rainfall is about 17 inches but last year it was less than nine and, in spite of the drizzle, this year will not be much better.

By Wednesday morning the rain has stopped and although there are clouds in the sky they are high and unthreatening. The wind, still cold and quite strong, seems to have blown away the rain they badly needed.

There are eight people employed on East Bungaree and although they are on different parts of the property today they all keep an eye on the sky. Yesterday's rain and the gloominess of the sky had made them hope that they might at last be in the middle of a slow-moving weather system that could produce rain in a significant quantity. It looks less likely today, but there is still hope that the clouds might build up again after these sunny breaks have passed.

At South Booborowie, Brian Jamieson unloads the motorbike from the ute and calls his dog as he rides into the paddock. He brought the weaners down from Old Belcunda by truck yesterday and left them here to enjoy the best feed they have ever seen. Soon they will forget the hardships of their childhood, but meanwhile they need a little help. There is a water trough in one corner of the paddock and Brian Jamieson now musters the sheep and drives them towards it. It is a big paddock and the sheep, unfamiliar with it, might take days to find it by themselves. So he will put them to this trough each day until he is sure they know where it is.

In the rambling 1930s homestead at Belcunda, Trevor Burton, the new manager of East Bungaree, gets up from his desk in the small office and hopes he can escape before the phone rings again. There is much to do — files to read, people to talk to, work to plan, letters to write — but much of it can be done in the evening after the kids have gone to bed. Now there are things to do that can only be done during the day and they must take priority over the paperwork.

Born in 1945 at Cleve in South Australia, Trevor Burton left school when he was fourteen to work on the family farm. When his father died three years later he took over the running of the place and with it a responsibility that would have daunted many people twice his age. Five years later,

might come to an end. With nobody in the family to continue the work, the property was offered for sale by tender and it seemed almost inevitable that this famous property would be subdivided into smaller farms and the stud dispersed.

This prospect worried Charles O'Connor and the staff of East Bungaree, as well as sheep men everywhere who knew of its achievements. And, surprisingly, it also worried a remarkable industrialist in Adelaide called Bill Wylie.

Born in 1906, Bill Wylie had spent much of his life making a sizeable fortune from car components in general and shock absorbers in particular. But he had done much more than that. An energetic man, in 1936 he had bought 53,000 acres of virgin bush near Kieth and had succeeded in turning it into a model sheep and cattle property, called Kangaringa, which he still owned.

Industrialist by chance and grazier by choice, he knew the importance of East Bungaree. 'There were years of breeding there which you simply could not replace,' he said, 'and I wanted to see it continue.' So, having paid $2½ million, he became the new owner of East Bungaree and immediately announced that his intention was to continue the development of the stud and to maintain all its traditions.

Today, East Bungaree consists of four separate properties, none of which were part of the original East Bungaree which Hawker drew by lot in 1906, although most were bought by him in later years.

The headquarters of the stud consists of 5,753 acres of rolling, park-like country called Belcunda which is situated

Trevor Burton, manager of East Bungaree.

The station area at Belcunda.

Lou Steer with an outstanding East Bungaree ram.

Weaners on good country at South Booborowie.

Overseer Brian Jamieson yarding sheep at Old Belcunda.

when still only twenty-two, he had successfully put his younger brother through school and, with some of the responsibility discharged, he went as overseer to Wittalocka, near Kieth.

In 1975 he was appointed manager of Mercowie Stud at Redhill and two years later left to join a stock and station agent in his home town of Cleve. It was in May 1979 that Charles O'Connor invited him to become assistant manager of East Bungaree. In spite of the uncertainty created by the death of Stanley Hawker a few months earlier, he had no hesitation in accepting. The chance of working with Charles O'Connor was too good to miss, even though it might be short-lived.

When O'Connor retired in June 1983, after spending twenty-nine years at East Bungaree, Trevor Burton took over as manager at the age of thirty-eight. 'Charles O'Connor *was* East Bungaree whilst he was here, and the success of this property was his greatest achievement. I want to continue his work and maintain the high standards that he set.'

He looks at the sky as he walks past the stone woolshed to the ute. The clouds cover more of the sky now but they are still high. There will be no rain today, but if they continue to build up there might be some tomorrow.

The following day it starts to rain again, and then it clears. But not for long. Above the distant hills the sky is heavy with indigo clouds which build layer upon layer to make a mass of changing tones that dominate the landscape. They move across the sky and as they approach the rain comes. It becomes heavier and puddles start to form on the dirt roads that crisscross the property. It is good soaking rain, the best they have seen for two years, and a few hours later it is at last doing some good.

'Has this broken the drought?'

'Not yet. It will need more than that, but we might be on the way.'

It is still raining the following day and men and vehicles can no longer move freely about the paddocks. On Belcunda, around the hills from the homestead, Harry Jaquet slides across the mud as he drives out to feed the sheep. They are waiting for him, standing in soddenness along the slope of the bare hillside, now gently running with water. He moves a lever in the cab and checks his watch as the hopper on the vehicle starts to lay a measured line of feed on the ground behind him. The sheep walk to the feed and stand shoulder to shoulder on either side of the line. Others, coming from further away, join on so that by the time Harry Jaquet completes his delivery there is a double line of feeding sheep stretching for a hundred metres in this empty, rainswept paddock.

Over at North Booborowie Trevor Burton is standing in the rain looking at the flock rams. They are the workers that will be bought by commercial buyers to improve their flocks and they make up much of the business at East Bungaree. They sell about 1,700 each year and most go to the arid areas of South Australia and Western Australia, as well as some to other States. Satisfied that they are holding their condition, he climbs into the ute and drives through the property to a white building which stands alone on the side of a small hill.

This is the temple of East Bungaree and it is dedicated to turning out magnificent rams that are as close to the state of the art as they can get. It is very close, for this is a traditional stud that has always insisted that rams should, above all else, look impressive. And they do.

Inside, the ram shed is divided into four sections which run the full width of the building. The rails are of smooth timber and overhead the rafters of the roof project an ecclesiastical pattern in the subdued light. There are narrow feeding troughs along each wall so that here the rams can dine in grill-room comfort instead of having to wait for the soup-kitchen in the paddock.

They eat most of the time, silently and thoughtfully. When they stand side by side with their jaws moving in a curious circular movement they look like characters from some woolly West Side Story wondering whether to put the boot in. But when they turn away it is with the dignity of be-wigged, pre-revolution aristocrats who know true power.

They are huge. Most of them weigh nearly 140 kilograms, considerably more than a grown man, and their massive curved horns look as if they could turn over a tractor. But although they occasionally push each other about, perhaps when one of them behaves in a manner not fitting for a temple, they are docile animals and easily handled.

For Lou Steer, who is in charge of them, it is just as well. Kneeling beside one ram, he is gently clipping the wool which surrounds its face. He is using scissors, not shears or blades, and he is working with utmost precision, turning the ram's head from time to time to make sure that the wool is perfectly balanced on each side. Satisfied, he takes a small file and gently rubs the tip of one of the horns, checking again to make sure that it is exactly the same length as the other. He stands back to look at the animal for a moment, then lets it go and starts on the next.

He is preparing two teams of rams for auction, and another team for showing. The auctions, at Tumby Bay and Naracoorte, will be held in a few weeks time and already a number of buyers have come to this shed at East Bungaree to preview the rams that will be sold there. It gives them a chance to inspect them more thoroughly than will be possible at the sale, and to work out the strategy of their bidding.

The show rams are being prepared for the show at Crystal Brook, one of the most important in South Australia and a crucial lead-up to the annual show at Adelaide. Although many famous studs have withdrawn from shows, they are still very important to East Bungaree and much time and effort is put into them.

They are vital to us,' says Trevor Burton. 'It's not the sales so much as the chance to compare our rams against those from similar studs breeding similar sheep. We can see how much progress we are making and it gives an early warning if that progress starts to slip behind our competitors'.'

The Crystal Brook Show, 1982. On the left, Charles O'Connor, who retired in 1983 after many years as manager of East Bungaree, with the Grand Champion; in the centre, Lou Steer with the Reserve Champion; and on the right Trevor Burton with the Medium Wool Champion.

Recently, they have not been slipping behind at all. At Crystal Brook in 1982 they had the Grand Champion ram, the Champion Strong Wool ram, the Champion Medium Wool ram, the Reserve Grand Champion, and the Reserve Champion Strong Wool — and came back with enough ribbons to farewell a liner. In the same year an East Bungaree ram fetched $41,000 at the Adelaide sales and the team of sixteen rams brought a total of $90,900 in little more than half an hour's spirited bidding.

In the shed, Trevor Burton and Lou Steer have lined up the Tumby Bay sales team and are looking at them with muted admiration which they are reluctant to show but which they can barely conceal. Then the rams turn away and clatter on the slatted floor as they walk to the side of the shed for more feed.

It is dark by the time Trevor Burton leaves the shed to return to the homestead at Belcunda, and it is still raining. The countryside is sodden now. The headlights sweep through the mists of falling rain and the wheels throw sheets of water over the fences on either side of the road. The rain is constant, generous and exactly what they wanted.

He parks the ute in a shed near the homestead and walks across to check the rain gauge. Others might run, or not even bother, but for him the feel of rain is a luxury he has not enjoyed for months and he makes the most of it now. He looks at the gauge.

'How much?'

'Just over two inches since Tuesday.'

'Does that break the drought?'

'Yes, it should be over now. Let's go in and have a beer.'

AUSTRALIAN BEEF

There were no cattle of any kind in Australia before the start of European settlement. The first to arrive were brought by the First Fleet settlers and had been bought at the Cape of Good Hope during the voyage from England. Unfortunately a number were lost in the first few months of the settlement through the negligence of the convict who was in charge of them. They were found several years later on good land to the south-west of Sydney where they had developed into a substantial herd. Meanwhile those that remained in the settlement were kept for milk and as draught animals. It was a long time before there were enough to kill for food, even though the settlement was often close to starvation in those early years.

Other beasts were brought to the colony in the next few years. Most came from England but a few Zebus were also brought from India. Soon, however, cattle that had been brought from England multiplied and the Zebus and Cape cattle died out.

It is not surprising that the early 'settlers preferred their familiar English cattle. As with other aspects of agriculture, it was some time before those early European settlers realised that what was successful in Britain, was not always successful in Australia.

For many years the cattle that arrived were fairly undistinguished and little attention was paid to pedigree or quality. Then in 1822 Patrick Wood of Tasmania imported the first black cattle. Although they were described as Black Fifeshires they were probably examples of the breed that was later called Aberdeen Angus. Little is known of them and it was not until 1858 that the first authentic Aberdeen Angus arrived, in the shape of four bulls which, by coincidence, were also landed in Tasmania.

Tasmania also saw the introduction of Herefords to Australia when the Cressy Company landed a shipment of these cattle at Hobart in 1826. They continued to import these cattle for many years and in 1854 landed an outstanding bull called Cronstadt which still appears in many pedigrees today.

Meanwhile, the first Shorthorns had also arrived. They were brought out in 1825 to stock land that had been granted to their English owner and were eventually settled on 20,000 acres that was taken up on his behalf in the Hunter Valley. Only a few months after their arrival another shipment reached Sydney, this time on behalf of the newly formed Australian Agricultural Company. These animals were regarded as superior to any that had been seen in the colony so far. They were managed well and by 1828 the Company had nearly 1,500 head of cattle on its extensive lands in northern New South Wales.

But the A.A. Company was an exception. Most cattle at that time were run on comparatively small landholdings by squatters who were more than willing to change from cattle to sheep and back again as the market demanded. The boom that was brought about by the gold rushes encouraged many to increase their herds, but only until prices fell. There was little long term commitment and no one of the stature of

Chasing cattle in 1878. REPRODUCED BY PERMISSION OF THE NATIONAL LIBRARY OF AUSTRALIA.

Macarthur to provide the kind of encouragement he gave to wool growers.

Cattle could be grown only for the Australian market. Without the means of exporting, it was clearly foolish to build large herds as not only would they exceed demand but would also make it much more difficult to change to sheep should the market make that necessary.

The aftermath of the gold rushes left a domestic population that had grown dramatically in the past decade and the demand for meat was much greater and also more stable. With some of the uncertainties removed, beef production could now start to develop as an industry in its own right.

With this came an urgent need for new land, as by now most of the fertile land in the east of the country had long been occupied. The South Australian government commissioned explorer John Stuart to find new land in the north of that State. He succeeded in opening new areas that were soon settled even though they were very isolated. His later journeys through the centre of Australia, which culminated in his successful crossing of the continent to its north coast, were even more important for they revealed land in enormous quantity.

300 Shorthorns, the foundation of a station that has been famous ever since. It was the very stuff of Australian legend.

Less dramatic perhaps, but even more significant, was the first shipment of frozen beef and mutton to England in 1879. People had been experimenting with refrigeration for years, but even after their success the problems of installing it on ships were still immense. Indeed, problems remained even after this first shipment and it was many years before they were finally solved. Nevertheless, this shipment was of untold importance. For the first time Australian meat producers could supply overseas markets and were no longer restricted to the demands of the Australian public.

This led to the settlement of more land for cattle grazing and the previously unattractive areas of the Cape York Peninsula and the Barkly Tableland were now used for grazing for the first time. Largely because of the increased demand brought about by refrigeration, or at least the promise of it, Queensland emerged as the main producer of beef in the 1890s.

But by then some of the excitement had gone out of the land boom, especially in northern Australia. With little knowledge and no control, cattle numbers had been allowed to increase in the good years until they had gone far beyond what the country could carry in poorer seasons. Then, the cattle would concentrate around natural water and the overgrazing that occurred resulted in loss of pasture followed by serious soil erosion — so much so that the carrying capacity of many areas was lower in the 1960s than when they were first settled. In the harshness of the outback the delicacy of the environment was far from obvious and much damage had been done by the time it was recognised.

Although speculators usually made a profit from the land, graziers were less successful. There is little evidence of any of them making much profit in the last thirty years of the nineteenth century. By the time the colonies were joined in Federation in 1901 the optimism of northern Australia had died.

Nor, in spite of promises, was there much change in the early years of Federation. An attempt to open new markets in countries to the north of Australia was not successful. When the meatworks in Darwin closed in 1920 it triggered off a series of depressions that were to have an important effect on land use in the north.

Low prices drove many off their properties which were then taken over by those who were financially stronger. Pastoral companies were already playing a significant role and now became even more important. With little export business going through Darwin, pastoralists had to overland their cattle to Queensland, where they were held on fattening properties before being sent to overseas markets.

The larger companies were able to acquire chains of properties on the major stock routes and these, along with the fattening properties they bought in Queensland, gave them a very big advantage. They could now move cattle with relative ease from one property to the next along the stock routes.

His glowing reports, together with the completion of the Overland Telegraph Line in 1872, led to considerable interest in this part of the continent. By the early 1880s there was a spectacular boom as people rushed to secure pastoral leases, often for land they had not even seen. By 1883 more than ninety per cent of the Northern Territory's 1¼ million square kilometres was held under application for leases, considerably more than was held even in 1969. The enthusiasm was real enough, though, and the area already carried about 61,000 head of cattle.

Getting them there was an incredibly difficult job. For example, in 1882 William Osmond and John Panton bought 4,000 head in Queensland and overlanded them to the Ord River. Here they established the first station in the Kimberleys in 1884.

Perhaps even more remarkable were the Macdonald brothers, who left Goulburn in New South Wales with a mob of cattle for this new area. Most of the mob died and they had to buy more in Queensland, but many of these died from redwater fever after crossing tick-infested country. In the end only one brother survived. Charles Macdonald arrived in 1885 to take up Fossil Down with the surviving

For the smaller grazier this was much more difficult. Unless he was able to reach the southern railway to Adelaide he was faced with the task of overlanding his cattle anything up to 1,200 kilometres into Queensland. The season determined whether he could complete this journey, or even start it, and market prices determined whether it was worthwhile even if he succeeded.

Consequently most tried to sell their cattle to another station on the way, usually a company station that could then move them with little risk to their next station along the route. So the small graziers came to rely more and more on these company stations and in turn the companies became more and more important. The 1920s saw the establishment of many absentee owners, many of whom were based overseas and whose stations were run by managers for maximum profit.

Even so, many companies saw little return for their very considerable investments and until the start of the Second World War the beef industry in the north of Australia was rarely far from bankruptcy. The war brought much relief and prices started to climb steadily as a result of overseas demand combined with a good domestic market which was increased by the growing number of American troops that passed through Australia. Prospects continued to improve after the war with the agreement with Britain and the opening of the American market.

As with other aspects of pastoral Australia, technology now played an increasingly important part. Many areas that had previously been unsuitable for cattle could now be improved by the use of Townsville stylo. When top-dressed with superphosphate it produced a vastly improved pasture which transformed marginal land. But perhaps even greater benefits of technology came in the new approach to cattle breeding.

In the south of the country British cattle did fairly well, although their success was often the result of better soil fertility and pasture growth rather than their ability to adapt to the country. In the north, however, it was a different story. There the extremes of heat and humidity and tropical parasites, such as cattle tick and buffalo fly, presented an environment that was very different.

Although the first cattle taken into these areas were British Shorthorns, it had long been realised that they did not thrive there. So agricultural scientists looked for other breeds that might be more successful, and the most obvious was the Zebu.

The first important introduction of Zebus were Brahmans from America, together with a number of others brought from Pakistan by the CSIRO in 1956. Soon there were many new breeds, all of which had been obtained by crossing Zebus with various British breeds. Outstanding were the Santa Gertrudis (Shorthorn and Brahman), and the Belmont Red (Africander, Hereford and Shorthorn) which was bred by the CSIRO at Rockhampton for commercial use.

Most of these new breeds combined the more placid temperament of British breeds with the heat tolerance and

tick resistance of the Zebus and they were readily accepted in northern Australia. In recent times they have also found increasing acceptance in southern areas, where the benefits are less obvious but still desirable.

In Australia most cattle stations breed their own herds and raise them for sale as vealers, yearlings or older cattle. Some concentrate on fattening store cattle that have been bought for the purpose, and some do both.

In eastern Queensland cattle are often sold through sale yards. This gives access to a number of competitive buyers but it is only practical if the property is close to a sale yard and if the number of cattle being sold is not too great. In the more isolated areas to the north and west, however, where the climate might reduce the working season to six months, the numbers being turned off are considerable. Because of this, most cattle in these areas are sold privately to the nearest meatworks.

Moving cattle to the meatworks is a costly exercise, as it can involve distances of several hundred kilometres. The development of all-weather beef roads in isolated areas has done much to simplify this. Although it is still expensive, large numbers of cattle are regularly transported over these large distances by road trains.

It might seem that in the process much of Australian legend has disappeared. But it has not, for in those remote parts of Australia legends are a long time dying, and when they do they are easily replaced by others. Technology has changed the way of life on the stations, but it has done little to change the men who work them. They are Australian cattlemen, and they are a rare breed.

VICTORIA RIVER DOWNS

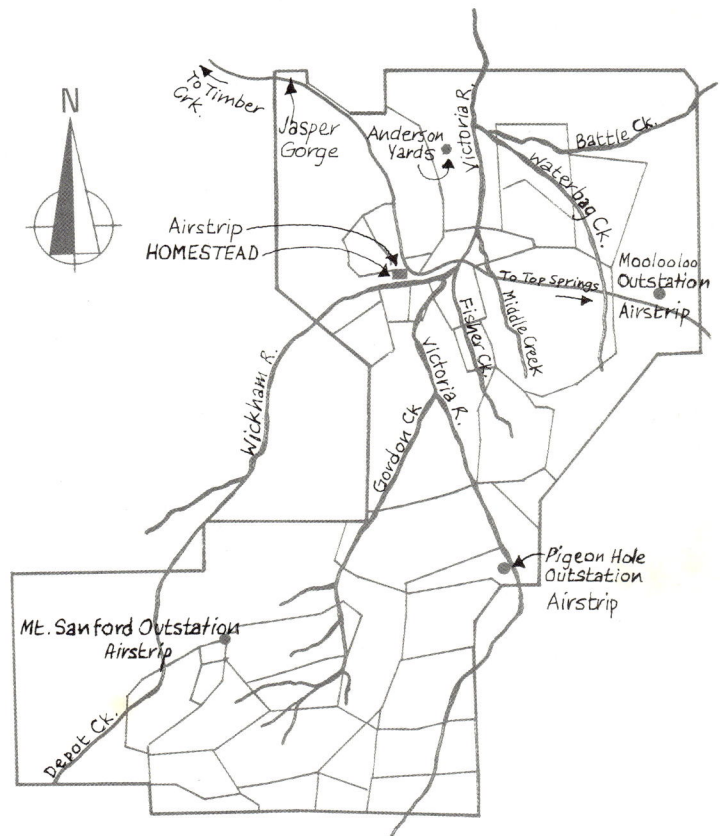

It is early morning and the air is still and not yet warm. Sunlight slants through the trees on the banks of the Wickham River and wallabies creep out from the scrub to drink from the mirror-still water, tinged yellow with the early light. There is hardly a sound. A few birds make experimental calls to each other from the frangipani trees and in the distance a dog barks once and is quiet again. It is a peaceful landscape waiting to come to life for another day.

A helicopter starts up with a clattering roar. The wallabies disappear back into the scrub, the birds fly off in confusion and the dog starts barking again. The roar gets louder and some of the unevenness goes out of it. Then another starts up, and another.

In front of the hangar, where the helicopters are sitting, the noise is shattering and brutal. Smoke comes from the exhausts and is blown away then the motors settle down and the smoke is replaced by clear hot air that shimmers briefly before it too is swirled away. The sun glints on the rotor blades so that they flicker like an old movie, whilst below them the glass bubble is streaked with colourful reflections.

The motors rev up, the blades go faster so that the flickering disappears and they become a continuous blur, and one by one the machines slowly lift themselves from the ground. Nose down, they turn and slide away over the buildings that surround the hangar and people still in bed curse them as they go. It is the start of another day at Victoria River Downs.

Victoria River Downs is one of the most famous cattle stations in Australia. Huge and remote, it combines all the romance, toughness and beauty of the outback — a self-contained world in the middle of Australia that few ever see.

It is 376 kilometres from the small Northern Territory town of Katherine over a road which is mostly good bitumen. In the winter tourists travel along it in their thousands, to and from the West Australian border and the Kimberleys beyond, but few ever notice the dirt road that turns south near the small settlement of Timber Creek, and even fewer turn on to it. This is not a good road for tourists, nor does it have to be, but it passes through magnificent scenery of rocky hills and scrub on its way to Victoria River Downs, the 'Big Run'.

Although not the biggest cattle station in Australia, Victoria River Downs is a huge property covering more than 12,000 square kilometres. From the limestone country of the northern boundary, near the spectacular Jasper Gorge, it is about 150 kilometres to the southern boundary near Wave Hill, where the country is semi-desert and covered with spinifex. At its widest, the east and west boundaries are more than 100 kilometres apart. The eastern border is hilly basalt country which turns into more open downs land further west. There, the spinifex gives way to Flinders and Mitchell grass which provides good feed all the way to the more rugged country in the north-east.

It is harsh country, almost barren — country you could die in if you were not an experienced bushman. But most of it is good cattle country and always has been. The paradox is that in spite of this Victoria River Downs has broken the heart of more than one of its owners in its hundred-year history.

It is was on 18 October 1839 that John Wickham, commander of the *Beagle*, discovered the mouth of the Victoria

Gwen McAntee sorting mail in the post office at Victoria River Downs.

River. He had been sent by the Admiralty to fill in the gaps that then existed on the charts of north-western Australia. But although he and his men spent a considerable time exploring this big river, they were unable to make their way very far up it.

It was not explored again until 1855, when Augustus Gregory landed at the mouth of the river at the start of an epic journey of exploration that was to take him across the continent to the Queensland coast. Gregory and his party succeeded in exploring the whole of the Victoria River and also discovered and named the Wickham, a tributary which runs alongside the homestead at Victoria River Downs.

Although many plans were drawn up to settle the area it was not until December 1879 that Charles Fisher and Morris Lyons made the first application for a pastoral lease. Fisher was then one of the richest men in Australia and Lyons, a Melbourne magistrate, was also a shrewd businessman. They owned another property called Glencoe and it was from there that the first stock was taken to Victoria River Downs. They gave the job to Nat 'Bluey' Buchanan, a legendary drover of his day who succeeded in arriving at Victoria River Downs in May 1883 with a mob of 20,000 Shorthorns.

Having stocked their vast new property, Fisher and Lyons soon found they had overreached themselves and they

offered Victoria River Downs for sale. There were no takers and in 1886 the firm of R. Goldsbrough & Co. secured a mortgage over it and saved Fisher from a financial ruin that would have had repercussions throughout much of the Australian business world.

The following year Goldsbroughs sold the property to the North Australia and Northern Territory Company, which had been formed in England for the purpose of buying it, for the remarkable sum of £300,000. It was complicated deal that involved the payment of cash, shares and debentures. Unfortunately not all these obligations were met by the new company and in 1888 the combined firm of Goldsbrough Mort started legal action. The case was fought in the courts of England and South Australia until both parties realised that costs were becoming ruinous and decided to come to terms. As a result, Victoria River Downs passed into the hands of Goldsbrough Mort.

The first manager resigned that year and the 'Gulf Hero', Jack Watson, became the property's second manager. He got his name following an incident in the Gulf of Carpentaria when he had dived from a ship to save a sailor who was being mauled by a shark. Armed with only a sheath knife, he repeatedly attacked the shark until it turned away.

He was equally fearless as manager of Victoria River Downs and often challenged people to join him in jumping a horse into the Wickham River from a ten-metre-high bank. Not many did. To Goldsbrough Mort, however, he had some infuriating weaknesses, one of which was his refusal to keep them informed about the property they owned. One report he sent, typical of many, said simply: 'Water, pasturage and Stock. Same as last month.'

Watson was drowned whilst swimming stock across the Katherine River in 1896. By then, Victoria River Downs was running a herd of 30,000 Shorthorns on their 20,000 square kilometres. But the company had received a bad mauling in the financial collapse of the Nineties and that year they offered Victoria River Downs for sale at a price of £75,000. They were too optimistic for such depressed times. It was four years before they found a buyer who would take the place off their hands, and then for the much more modest sum of £27,500.

The buyer was a syndicate of three men who knew a great deal about running cattle stations. They were explorer Alexander Forrest; his associate Isadore Emanuel, who had pioneered the cattle country of the West Kimberley; and cattle-king Sidney Kidman. Forrest died the following year but the syndicate continued and set about improving the property. They built yards and put up fences. The new manager, James Ronan, who had been one of Kidman's cattle buyers in Queensland, tried to improve the management of the stock.

But having consolidated four additional leases, Victoria River Downs now covered nearly 24,000 square kilometres and it was simply too big to run. Cattle thieves roamed across the unfenced property and Ronan turned a blind eye to them in the hope that their presence in remoter parts might deter the natives who were killing cattle, as well as each other and an occasional European. The property had cattle to spare for a few thieves (the books showed a stock of 34,000 head but there were probably another 60,000 that had never been branded). In 1903 Ronan was fired and Dick Townshend, who had run the Wyndham branch of Forrest and Emanuel, took over as manager.

During the first four years the new owners turned off a total of 14,000 bullocks. Most were sent by sea to Perth, where the syndicate also controlled the wholesaling of meat to the goldfields at Kalgoorlie. It was very profitable, but as that market declined the property declined with it and it was sold once more.

In 1909 Bovril Australian Estates, a subsidiary of Bovril Ltd in England, paid the syndicate £180,000 for Victoria River Downs and a small station called Carlton Hill near Wyndham. This was the biggest land sale so far completed in Australian history, and probably produced one of the biggest capital gains as well.

Bovril, who had many overseas properties, had examined the Victoria River area and the purchase of Victoria River Downs fitted in well with their plans. Unfortunately it was the start of a disaster that was to last for fifteen years.

The problem, and it was one that affected nearly all northern cattle stations at that time, was that there were no local meatworks. Although there had been many plans to establish them at Wyndham and Darwin, nothing had ever happened. In 1917 Bovril joined with Vesteys to build a works at Darwin, but high running costs and a series of industrial disputes forced it to close three years later.

At 9¼ million acres, Victoria River Downs was then the biggest cattle property in the British Empire, but it was barely paying its way. It was, as the chairman of Bovril said, possible to ride for a week in a straight line without leaving the property, and to see nothing but cattle whose destiny was to be converted into Bovril. The trouble was the conversion.

Although a new meatworks at Wyndham gave some relief (indeed in 1930 a third of all the cattle killed there came from Victoria River Downs), the Depression made life even more difficult. In 1934, with over 170,000 head of cattle on the property, the gross profit was less than £20,000 and the following year, unable to meet the government's demand for more investment, over 3,000 square kilometres were lost through resumption.

The Second World War brought with it a contract to supply the army in Darwin with 400 head each month, most of which were processed through a new meatworks which had been built near Katherine by Vesteys. But Bovril resented the control that Vesteys could exert and in 1948 they decided to build a works of their own in the same town. This was yet another disaster and when it failed in 1950 the financial collapse of Bovril seemed unavoidable. In 1955 Bovril sold Victoria River Downs to an Australian millionaire, William Buckland, and with a sense of relief turned their attention back to their more profitable ventures in the Argentine.

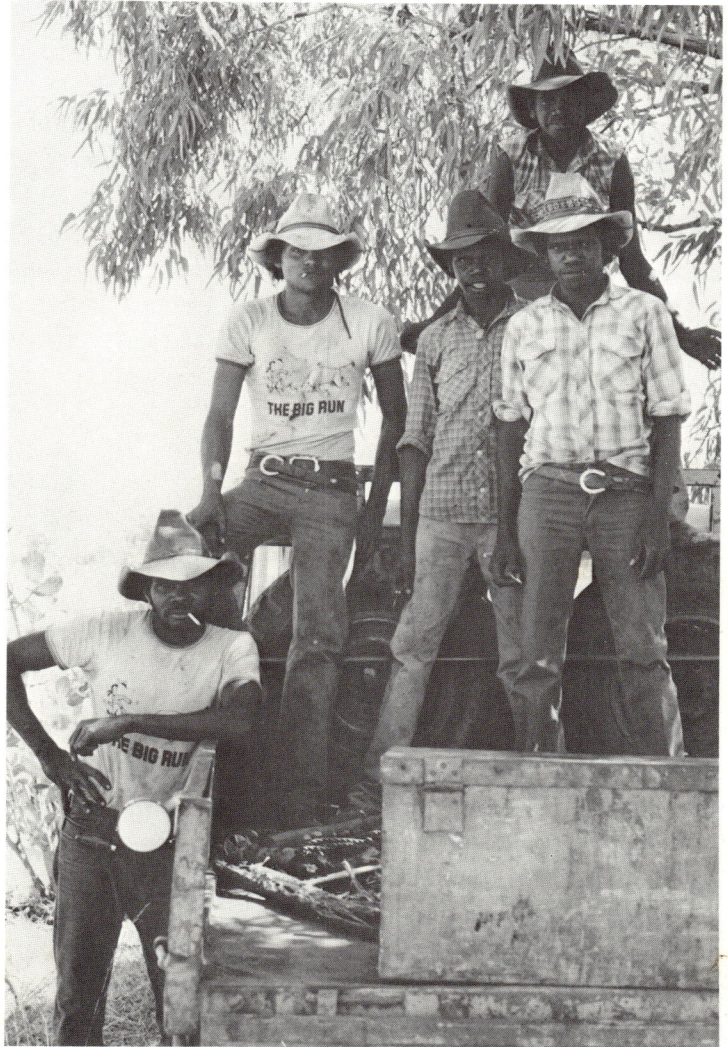

Aboriginal stockmen at Victoria River Downs.

Buckland, who had made a fortune in service stations and car accessories, liked the idea of being a pastoralist. But his expertise was in the cities, where he had a newsworthy reputation for being able to move vast sums of money into profitable ventures. During the time he owned Victoria River Downs he never once visited it.

His skill as a businessman was real enough, however, and in 1960 he pulled off one of his greatest deals when he sold the property to L. J. Hooker Investment Corporation. Once again it was the biggest land deal in Australian history and it made the Hooker Corporation one of the biggest landowners in the country.

Hookers soon realised that the property they had bought was not in the best of condition. The stock was virtually out of control and improvements were urgently needed. They started by turning off a large number of cattle at almost any price and then spent thousands of dollars repairing fences and building new ones, especially in the southern part of the property which could be worked more easily than the hillier country to the north.

They built 24 new drafting yards and sank 46 bores to carry the property through a drought that did not end until 1967. Then, in their first good season, they bought a herd of Brahman breeders to improve the herd which, apart from some Hereford bulls which Bovril had introduced in 1927, had remained virtually unchanged since the days of Fisher and Lyons.

These improvements were a long and expensive business and went on for years. There are now 418 kilometres of boundary fencing and nearly 2,500 kilometres of internal fencing, together with 4 registered and 12 unregistered airstrips. There are 70 bores to supplement the natural water of the Victoria and Wickham Rivers and there are a hundred paddocks, most of which cover about 32,000 acres. There are 90,000 head of cattle consisting of just about everything between Shorthorns and Brahmans and 15,000 head are turned off every year to the meatworks at Wyndham, which is also owned by Hookers. When Victoria River Downs celebrated its hundredth birthday in May 1983, Hookers could boast that the property had shown more profit and development in the last twenty years than it had in the whole of the first eighty.

But for manager Gilbert McAntee, that development is not finished yet. Parts of the property are still not fenced and watered and in the northern area they do not have complete control of the herd. Although the cattle are free of brucellosis, there is a small incidence of tuberculosis. They are eradicating this by mustering the northern area as cleanly as the scrub country allows, then turning off the mature beasts and TB testing the weaners before moving them to more controlled parts of the property. He hopes to have control over all parts of the property in about three years and then he will be able to start a breeding programme to improve the herd.

Gil McAntee was born at Wondai in Queensland in 1938. He left school when only fourteen and went to work on the family dairy farm, where he saw at first-hand the hardships caused by drought. When he was sixteen he moved to another farm as a horsebreaker and stayed there for two years before moving to Westgrove Station north of Injune in Queensland. When he was twenty-five he spent a year with an earth-moving company before joining Boam Downs as a ringer. In 1966 he became manager of a mixed farm which ran sheep and cattle. Shortly afterwards Gil was able to buy the old family farm and a 250-acre block which adjoined it. In 1971 he was appointed assistant manager at Victoria River Downs and took over as manager in 1974.

His career has made him an intuitive manager. Competent and quietly spoken, he invites respect instead of demanding it. He has had no difficulty fitting in with a big organisation. His job is running a cattle property, and he knows more than enough to do that. Once, when he was assistant manager, he wanted some cattle moved from one part of the property to another. The head stockman, a man of many years' experience, told him that it could not be done at that time of year. So he collected his two children, who were home from school, and, with a couple of Aboriginal ringers, they moved the cattle over the weekend. The head stockman left, and not many people have since challenged Gil McAntee's view of the possible.

Gil McAntee, manager of Victoria River Downs — the Big Run.

Mark McAntee, head stockman with the Centre Camp.

The official post office in the station area at Victoria River Downs.

The huge trailers of a road train reflected in the driving mirror.

The homestead, Victoria River Downs.

Running Victoria River Downs is a big job by any standard. With three outstations (Moolooloo, Pigeon Hole and Mount Sandford) there are sixty-five people on the payroll and the organisation has to be very professional if the place is to run smoothly.

Because of its isolation Victoria River Downs has to be virtually self-contained. Stores are ordered twice a year, once for the Wet season and a bigger order six months later for the working season. The orders are made out by Gil's wife Gwen for all stores — equipment, wire, tools, spare parts, ammunition, furniture, stationery, food, clothing — everything that the station will need for the next six months. Goods are delivered to a transport depot in Brisbane, where they are consolidated into one load for a road train. When it leaves Brisbane it will be carrying about 50 tonnes of stores and two drivers will take turn about for the three-day drive to Victoria River Downs.

Some of the stores are distributed to the workshops and outstations but clothing, food and other consumer goods are taken in as stock for the station store. This is a comprehensive shop which opens three afternoons a week for the benefit of station staff and their families and the Aborigines who live on the station or in the nearby settlement. It has a turnover of more than $350,000 a year.

Alongside the shop is a fully operational, and official, post office which handles mail and provides a counter service. Mail arrives on Wednesdays and Fridays by plane from Katherine and a station plane then takes it on to the outstations. It is not unusual for the incoming mail to consist of five hundred items and there might be as many again to be loaded on to the plane to be taken back to Katherine.

The station area, which even has street lighting, is big and impressive in its neatness. There are extensive lawns that are constantly watered from the Wickham River and which look refreshingly green in what is often arid country. The street in front of the large homestead on the bank of the river has also been turned into a lawn and the effect is not unlike a village green with buildings dotted around it.

The post office and store are at one end and behind them are the maintenance areas and workshops that keep the enormous amount of machinery in good order. There are massive graders which are used to repair the kilometres of roads after the Wet, and a road train that takes cattle to the meatworks in Wyndham. It is a twelve-year-old Mack R700 and, with its rigid trailer and three dog trailers, it is 49 metres long.

Across the green from the homestead is a clinic run by a full-time nursing sister, whilst further on is a school for station children run by the Northern Territory government. Nearby are houses for station staff, whilst in the middle of this huge complex is an area called the bull ring which serves as the children's playground. Beyond that are the single men's quarters and their air-conditioned dining room and further on, not far from the main airstrip, are the hangars of Heli-Muster, an independent company which operates from this base at Victoria River Downs.

Heli-Muster was started in 1976 with two helicopters and two years later moved on to the property to combine with the two station helicopters that were already in use there. It now has a fleet of twenty-two helicopters and seven fixed wing planes, together with two turbine helicopters which are used for survey work and geological exploration. The company has a variety of clients throughout the Kimberleys, the Northern Territory and the Gulf country of Queensland but a major part of its work is on Victoria River Downs itself, where the techniques of mustering by helicopter were first developed.

Now, four helicopters and five fixed wing aircraft are in permanent use on the property. They are flown by a staff of about nine pilots. Surprisingly perhaps, none of the helicopter pilots are Vietnam veterans. They are civilian-trained, often at considerable personal expense, and more than half of them have experience of mustering cattle on horseback, so that the techniques they have developed are an amalgamation of old and new.

Skilled ground support is essential for successful mustering by helicopter and this is now provided by the stock-camps. There are four stock-camps in the dry season at Victoria River Downs. One of them, the Centre camp under Gil's son Mark, is now waiting for the helicopters near the Victoria River. He and his Aboriginal stockmen spent the previous day laying dumps of Avgas drums along the route of the muster and taking down sections of fencing so that the helicopters will be able to push the mob from one paddock to the next on its way to the yards. Now he and his men are parked beside the first fuel dump waiting for the helicopters to land for refuelling.

The three helicopters that left the station in the early light have already been at work for a couple of hours, swooping from side to side as they flush beasts from the scrub and watercourses. Much of the country is heavily timbered, especially near the river, and some of these cattle have never been mustered before.

Having taken up residence in places almost impossible for a horseman to penetrate, there is enough good feed and permanent water for them to grow to maturity without them having to venture into more open country. Helicopter mustering is the only practical way of getting them out. The machines swoop over them and the clatter of the blades and the roar of the motor, which disturbed the tranquillity of the station area that morning, now make the beasts run from cover. At first they run in different directions but the pilots, often flying right behind them, start to collect them together in groups and combine these to form a small mob. When the mob is moving in the right direction the pilot will veer away to flush out more beasts to swell the numbers.

Mark McAntee stands by the ute watching them. Then as one of the helicopters leaves the others he calls to his men that the first machine is coming in to refuel. It slides round in a steep turn, then settles itself gently on to the ground beside the drums of fuel. The men have already put the hand pump into the drum and the machine is barely at rest before

Recently mustered cattle being held in portable yards.

one of them climbs up to put the end of the pipe into the fuel tank. Another man turns the handle of the pump whilst Mark has a quick conversation with the pilot under the whirring blades.

The stockman hands the pipe to a man on the ground and replaces the cap on the tank. The pilot adjusts his helmet and the men stand back as he speeds up the motor. The blades blur again and the machine lifts from the ground and turns away. It has taken no more than three minutes.

When they have refuelled all three helicopters the stockmen pack the gear into the vehicles and start to move to the next dump. For them it is a leisurely business and will be so for most of the day. The vehicles must stay behind the cattle, following slowly as the helicopters continue to sweep and dive. By the middle of the day, after refuelling the helicopters again, the men on the ground have not yet seen a single beast, although the pilots say they now have a mob of several hundred head. So far they have come about 6 kilometres and the remaining 10 kilometres to the yards will take the rest of the day.

With the helicopters airborne once more the stockmen light a fire, put the billy on, and open the tucker box on the back of one of the vehicles. In it is damper, cold tongue and other meat, and the remains of a 6-kilogram barramundi which the cook caught at their camp on the Victoria River the previous day. The men help themselves from the box, pour themselves some tea, and settle down in the shade of

the trees. Although it is April, the temperature is well into the 30s and it will be a long day. Mark rests easily with his men. He grew up with them on the station, went to school with most of them, and they know each other well.

As they finish their meal they are surprised to see a helicopter land nearby. It refuelled only an hour ago and they did not expect to see it again until the middle of the afternoon. Mark runs over and leans into the glass bubble. Shouting over the noise of the machine, the pilot explains that the position of the mob makes it possible to take a more direct route to the yards and so save time, but in doing so they will miss the gap in the fence. The other helicopters are holding the mob out of the way so that Mark and his men can make another gap for them.

As the pilot takes off Mark is already organising his men into one of the vehicles. They drive ahead for a kilometre or so, then bounce off the road to follow the fence. They stop about a kilometre from the river and the men jump out. Some quickly cut the strands of barbed wire and unclip it from the posts, whilst others start to coil it or pull the metal posts from the ground. By the time they stack the material out of sight there is not a trace of the fence. They have removed about 1,300 metres of it in fifteen minutes and as soon as the men have left, the helicopters will start moving the mob through.

By the middle of the afternoon the helicopters have refuelled for the last time and the men on the ground can

The end of the muster.

now see the dust of the mob as it moves slowly towards the Anderson yards. At five o'clock, when the mob is pushed through the gap in the fence into the last paddock, Mark McAntee can see the beasts for the first time. His role changes now. As the helicopters move the cattle across the open ground of the paddock, he drives up behind the mob and revs the motor to keep the cattle moving.

Two of the helicopters are moving a mob that has been separate until now and as they join together there are hundreds of beasts walking across the paddock, dust trailing behind them in a long, low cloud. Suddenly some beasts burst away and gallop off in different directions. The third helicopter swoops after them, turning and spinning as the cattle prop and change direction. By the time the pilot has turned them to head back to the mob the skids of the helicopter are practically touching the tail of the last beast.

This is a big paddock and it is half an hour before Mark can see the start of the wings — two long fences which converge like a funnel towards the gate of the yards. The mouth of the funnel is about 1½ kilometres from the yards and is several hundred metres wide, but this gap becomes progressively narrower as it approaches the gate. The fences are hung with hessian so that the cattle cannot see beyond them, leaving them no alternative but to walk forward into the narrowing neck. It is an old technique that was once used to catch brumbies and which has now been revived for use with helicopters.

As the cattle reach the mouth of the funnel they are forced into a tighter mob. Mark is only metres behind them now, revving the engine and blowing the horn to keep them moving whilst overhead the helicopters, perilously close together, fly from side to side in a tight ballet. They drop lower and lower as the cattle move along the funnel and the pace quickens in the more confined space.

The leading helicopter is now not much higher than the ute and dust comes over the vehicle like breaking surf. The noise of the motors and the clatter of the blades drown out the bellowing of the cattle as they are forced along the funnel. The leading beasts start to go through the gate into the yard and the helicopter darts from side to side in what seems like a decreasing spiral above the narrow gap. Immediately below, Mark keeps his hand on the horn as he coaxes the last of the mob through the gate. After many hours, and from different directions, men, cattle and machines have finally converged in a crescendo of noise and dust at the Anderson yards.

The stockmen leap from the back of the ute and run forward to close the gate as the pilots land the helicopters nearby. They have mustered about 700 head of cattle from three huge paddocks and yarded them, all without the use of a single horseman.

Mark and his men strengthen the gate with extra wire. Then they will return to their camp by the river, hoping that the cook has had another successful day. The pilots stamp out their cigarettes, pull on helmets once more, and get ready to go home.

It has been a long day and as they head back across country the light starts to fade and the red of the sunset flares into the sky. Darkness spreads slowly over the ground and their navigation lights wink at each other as they fly in a loose formation over kilometres of nothingness. Then in the distance some lights come on — a small cluster of pinpricks in an empty, darkening landscape. They are the lonely lights of Victoria River Downs — the Big Run.

MOVING A MOB OF CATTLE

Mobs of cattle are moved for a number of reasons. They might have to be tested for disease, calves might need to be branded, weaners separated from their mothers, or mature beasts shipped off to the meatworks. It is a frequent, almost constant, job on any cattle station and being able to move a mob of cattle correctly is one of the basic skills of a good stockman.

Techniques can vary depending on the size of the mob, the temperament of the cattle, the distance they have to be moved, the type of country to be crossed, and the number of men available. Moving a handful of docile cattle across a couple of open paddocks is relatively easy, but moving several hundred head of wild Territory cattle across rugged country for days on end is a different matter. Nevertheless, the techniques will have much in common.

Once the cattle have started to walk they will be formed into a moving column and the stockmen will take up the positions shown in the diagram. The most experienced man usually takes the lead. He sets the pace at which the mob will move and 'steers' it in the right direction. The pace is very important. Cattle usually take up the same position in the column each time, with the stronger beasts in the lead and the calves and older beasts at the back. The lead man has to make sure that the stronger beasts do not walk too quickly and he holds them at a pace that can be managed by the slower ones at the back.

The ability to move a mob of cattle is one of the basic skills of a good stockman.

The wing men ride on either side of the mob to contain it. They will deter beasts from dashing to freedom and ride after those that do and bring them back to the mob. The men on the points keep watch on the beasts behind the wing men, whilst the man on the tail guards the rear of the mob and keeps it moving so that the column does not become too extended. If there are no points, the tail man will ride from side to side to watch the mob behind the wing men.

The men must remain in formation. The wing men might be anywhere along the sides of the mob, but those riding behind must know where they are all the time. The aim is to keep the cattle moving slowly and quietly. If men shout to each other, or dash about unnecessarily, the cattle will become excited and be much more difficult to control.

There is an important psychological element present all the time. The man and his horse have to dominate the cattle, imposing themselves so that the cattle know they cannot run away or drop back. Each man has to out-think each beast and anticipate what it might do, so that by the time it starts to move the man is already in position to stop it. This combination of instinct and experience is vital in keeping the mob under control. Ideally, and given time, the cattle will develop a 'mob complex' and behave as a single unit rather than as a collection of individuals.

If a beast does break from the mob, it is the job of the nearest rider behind to round it up. The others will then

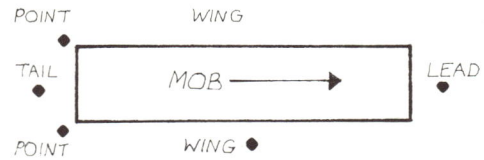

The positions taken by stockmen when they are moving a mob.

space themselves out to cover the gap he has left and will not join in the chase unless told to do so. There is a kind of etiquette, too. If one of the points rides behind the mob to chase beasts breaking from the other side, he is implying that the men on that side are not competent. This will be the cause of much bad feeling.

The mob is stopped by blocking it. The lead rider stops and turns to face the cattle. The front of the mob comes to a halt and the rest stop when they realise that those in front are no longer moving.

If the mob is to be halted for any length of time, perhaps for a meal break, the cattle will be allowed to graze freely in an area defined by the stockmen remaining on watch. This is called tailing. If the purpose of moving the cattle is to put them on good feed, they might be tailed all day and allowed to wander over quite a large area. Men then ride slowly round the perimeter of the mob like sentries, each covering a different section. It is then a simple matter to form the cattle into a column again when it is time to move on.

Rushes, the Australian term for stampedes, are mercifully rare but they do happen. The job then is to try to block the lead at any cost, regardless of personal safety.

ROCKLANDS

Two hundred kilometres west of Mount Isa in Queensland, and only a few kilometres from the Northern Territory border, is the small town of Camooweal. It has one main street and an airstrip which was put in during the war and which is nearly as long as the street. There is a post office, a hotel with a pet emu, and a couple of service stations. It is small and flat and if it was on the Hume Highway you would drive through it in a few minutes and forget it just as quickly.

But not many people drive through Camooweal without stopping. The next fuel is 200 kilometres away in either direction and there is nothing but empty country in between. So prudence, and perhaps hunger, make most people stop in Camooweal. There is usually a road train or two parked at the side of the road and cars are clustered round the service stations, if only for a few minutes.

If most people stop at Camooweal, not many stay there. And those that do are not tourists, they are the cattlemen who work the huge properties that spread across the Barkly Tableland and who know this small town as the 'Big Weal'.

I collect a cup of coffee in the roadhouse. Tourists are examining tea towels and kangaroo skins, whilst transport drivers roll cigarettes and study the racing pages.

'Can you tell me how to get to Rocklands?'

'Down that road.'

Two dirt roads meet alongside the building before they join the bitumen.

'Which one?'

'The one with the grid.'

'How far is it?'

'About six kilometres.'

I thought it would be further than that. Rocklands is a big station and I had assumed the homestead would be many kilometres away, surrounded by the endless flat country of the Tableland. Now it seems it is almost a suburb of Camooweal.

But it is not quite like that. For 25 kilometres on either side of Camooweal most of the land alongside the road is part of Rocklands. But it is only a small part.

Rocklands covers 7,770 square kilometres and the southern part completely surrounds this small town and its local reserve. Indeed, Rocklands is older than the town and for many years Camooweal was little more than an outstation that provided facilities for the drovers who brought mobs of cattle along the stock route. Now, mobs of cattle travel by road train and occasionally mobs of people travel the same road in air-conditioned buses. But the role of Camooweal has not changed much and it doesn't look very different either. The same is almost true of Rocklands.

William Landsborough was the first European to see this country. In August 1861 he was appointed by the Queensland government to lead an expedition in search of Burke and Wills. He went by sea to the Gulf of Carpentaria and set off in search of the missing pair. He found no trace of them but he did find this vast area of flat pastoral land which he called the Barkly Tableland. Indeed, it was suggested later

The photograph on the left shows a stock camp in 1946. On the right, a motorised stock camp two years later.

137

Preparing to test a mob of cattle for TB and brucellosis.

that he had perhaps been more interested in discovering good land than finding Burke and Wills. He denied it, but his journals certainly contained numerous references to the pastoral potential of the land he crossed.

Making his way to the south-west, Landsborough crossed a vast black soil plain watered by a substantial river which he named the Herbert. Its name was later changed to the Georgina after the wife of the Queensland Governor, and Herbert, who was only Premier, was given another northern river as consolation. On 23 December 1861 Landsborough found a huge lake, which he called Lake Mary, and a week later he carved a record of his visit on the trunk of a tree.

Although few doubted the potential of the Barkly Tableland, there was still plenty of good land to be taken up closer to the towns. It was not until 1884, more than twenty years after Landsborough's exercise in graffiti, that the first pastoralist moved in. His name was Sutherland and he brought with him a mob of 8,000 sheep which he had overlanded more than 2,000 kilometres. He established himself by Lake Mary and called the property Rocklands.

He was there for four years, after which he wrote sadly that the Barkly Tableland was not good sheep country, although he thought it might be suitable for cattle. He was correct on both counts and Rocklands, the first property to be settled on the Tableland, has run cattle ever since.

Rocklands is now owned by the Stanbroke Pastoral Company of Brisbane. Ninety-seven per cent of that company is owned by AMP. So thousands of policy holders throughout Australia own at least a small part of this cattle station, even if they are not aware of it.

The drive from Camooweal is easy when the road is dry, and frequently impossible when it is not. The road meanders pleasantly for a few kilometres and then runs alongside the station airstrip before turning into the homestead area at the side of the lake.

This is still the original homestead area and it has developed through a long process of evolution rather than by careful design. People have lived and worked here for a hundred years and if it lacks precision this is because they preferred it that way.

At one end is a group of modern houses used by station staff, but further along the buildings are older and closer together. On the right is a row of buildings which might have been intended to form a street that was never finished. In the middle of the row is the station store, which is now the focal point of the station. Built in 1891, it contains a tiny office built into part of the verandah.

Facing it across the open space is the men's quarters. This building might be as old, but it has not been there as long, having been bought from the Methodist Inland Mission in

Yarding a mob of cattle in one of the vast paddocks at Rocklands.

1946 after the earlier building had burnt down. Behind it, and almost hidden by it, stands the original homestead. Small and solid, it was build in 1888 and now serves as the jackeroos' quarters.

Alongside it is the present homestead. Built in 1960, it is only the third homestead to be built at Rocklands in a hundred years. The second was demolished to make room for this modern building. It is a rambling single-storey house with a big kitchen and comfortable living rooms opening on to a wide verandah. In the garden, Landsborough's tree stands preserved under a little tin roof not far from where it was growing when he carved his record on it.

To the north, the boundary is 128 kilometres from the homestead, whilst the southern boundary is some 32 kilometres south of Camooweal. To the west the property crosses the State border and extends into the Northern Territory. There are only 33 paddocks in these 7,770 square kilometres and most of them are huge. One is 32 kilometres long and between 8 and 11 kilometres wide. It covers about 76,000 acres and there are others nearly twice that size.

The country is flat and seemingly endless. There are a few patches of gidgee scrub to relieve the monotony and there are lines of trees along the watercourses but beyond, the horizon is ill-defined, shimmering in nothingness and so far away it doesn't matter.

It is dry country, but it grows a rich cover of Flinders and Mitchell grass and cattle do well. The paddocks have never been cleared or resown and the country looks pretty much as it did when Landsborough first saw it.

It has a rainfall of 14 inches a year which mostly falls during the summer months of the northern Wet. Then the rivers and creeks run high and the rich black soil turns gluey. After this seasonal rain the ground dries out and the creeks become a series of waterholes. Some are permanent but most disappear under the summer sun, searing from a cloudless sky for weeks on end. There are 42 bores pumping water from below the ground so that the stock can survive.

They run about 36,000 head of cattle at Rocklands and most of them are pure English Shorthorns. It is one of only two Shorthorn herds in the area and they thrive on the rich protein that is a feature of much of this country. There are also 5,500 Africanders, handsome red beasts that will thrive anywhere on the property but which spend most of their time in those areas not suitable for the Shorthorns. Africanders were introduced to Australia in the 1950s and this is probably the biggest herd in the country.

Rocklands turns off about 6,000 beasts a year and most are sent as store cattle to one of the company's fattening properties in the Channel country of south-west Queensland. They also run about 400 horses, most of them

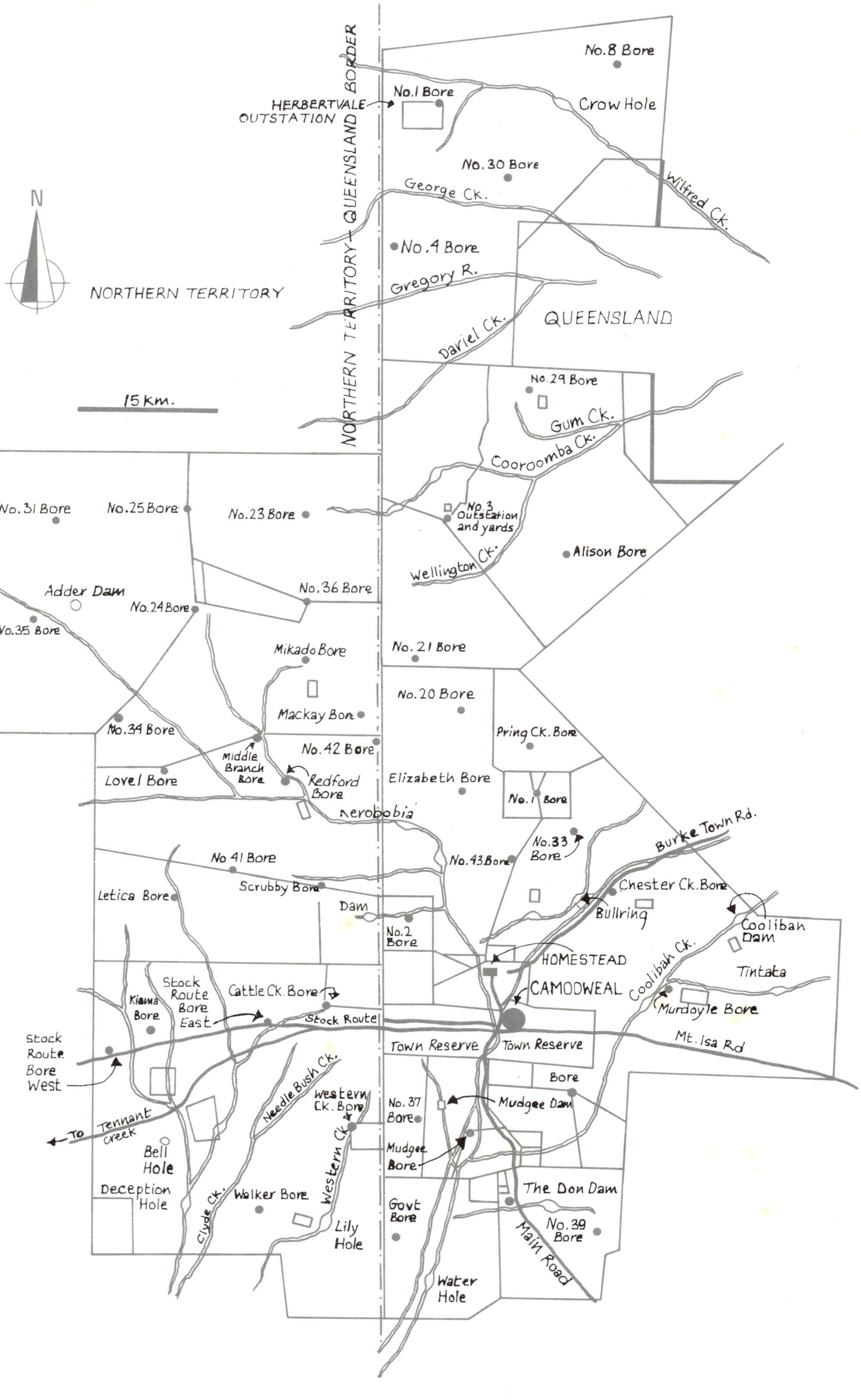

N

NORTHERN TERRITORY

NORTHERN TERRITORY - QUEENSLAND BORDER

QUEENSLAND

15 Km.

No. 8 Bore

No. 1 Bore

HERBERTVALE OUTSTATION

Crow Hole

No. 30 Bore

George Ck.

Wilfred Ck.

No. 4 Bore

Gregory R.

Dariel Ck.

No. 29 Bore

Gum Ck.

Cooroomba Ck.

No. 31 Bore

No. 25 Bore

No. 23 Bore

No. 3 Outstation and yards

Alison Bore

Adder Dam

No. 36 Bore

Wellington Ck.

No. 24 Bore

No. 35 Bore

Mikado Bore

No. 21 Bore

Mackay Bore

No. 20 Bore

No. 34 Bore

Pring Ck. Bore

No. 42 Bore

Middle Branch Bore

Redford Bore

Elizabeth Bore

Lovel Bore

No. 1 Bore

Nerobobia

No. 33 Bore

Burke Town Rd.

No 41 Bore

No. 43 Bore

Scrubby Bore

Chester Ck. Bore

Letica Bore

Bullring

Dam

Cooliban Ck.

Cooliban Dam

No. 2 Bore

HOMESTEAD

Tintata

Stock Route Bore East

Kiawa Bore

Cattle Ck. Bore

CAMOOWEAL

Murdoyle Bore

Stock Route

Stock Route Bore West

Town Reserve

Town Reserve

Mt. Isa Rd.

To Tennant Creek

Needle Bush Ck.

Western Ck. Bore

No. 37 Bore

Bore

Mudgee Dam

Bell Hole

Mudgee Bore

Western Ck.

Deception Hole

Clyde Ck.

Walker Bore

Govt Bore

The Don Dam

Lily Hole

No. 39 Bore

Main Road

Water Hole

Some of the old station buildings at Rocklands. Most were built in the early 1890s.

Ray Jansen, the manager of Rocklands.

thoroughbreds from their own registered sires. They are bred to work on the property, but when they do sell a few surplus horses there is never any shortage of buyers

Ray Jansen, the manager of Rocklands, is tall and wiry and is almost everyone's idea of what an outback cattleman should look like. Born at Collinsville, Queensland, in 1927, he left school when he was thirteen to become a butcher. He stayed at it for three years and then did a year's droving before becoming head stockman at Moray Downs, north of

Clermont. At twenty-three he became manager of a neighbouring property, Cumberland Downs, and then after more time droving he went to Gordon Downs in the Kimberleys as overseer.

This property was owned by Vesteys and shortly afterwards he became manager of Nicholson, another of their Kimberley properties. In 1956 he moved to Limbunya, an isolated property near Wave Hill, and stayed there for eight years before moving to Morestone near Camooweal. Two years later he left Vesteys and went as manager to Newcastle Waters in the Northern Territory. In 1972 he joined Stanbroke as manager of Stanbroke Station, south of Cloncurry, and in 1976 he took over as manager of Rocklands.

The property was in good condition, but since then there have been a number of changes. Most of these, as on many properties, have been brought about by changing economics and particularly by the increased cost of labour. Management has to be sharper and the property has to be run with fewer men than was once thought possible. There are now twenty-two people on the property, including three jackeroos.

The men have changed too. Although some are as good stockmen as ever rode across Australian stations, the standard has fallen in recent years. A good stockman needs an instinct for cattle that is then developed by experience. Some certainly have that instinct but often lack the broad experience that in earlier days would have come from droving.

Other, perhaps even more significant, changes have been brought about by the need to eradicate diseases in cattle. Two of these, tuberculosis and brucellosis, are the subject of a national eradication campaign.

Tuberculosis is the same disease that afflicts people and it can be passed on by contact with cattle. Brucellosis is a disease which leads to abortion in cattle. It too can be passed on to people, causing a recurring disease not unlike malaria.

Neither of these diseases is common in cattle but they are a danger, however small, to stockmen and meat workers. Further, a number of countries who import Australian meat, including the United States, have imposed a time limit by which Australian cattle have to be clear of them.

The eradication campaign has led to major changes in the way cattle are managed. For the programme to be successful every beast in a paddock needs to be mustered for testing. If any are left behind they might re-infect the rest when they are returned to the paddock and the diseases could continue even though the tests had revealed no presence of them.

On smaller properties this is not a major problem and in New South Wales and Victoria, for example, these diseases have virtually disappeared. But on the big properties of the north it is far from easy to muster all the cattle from one paddock, especially if the paddocks cover hundreds of square kilometres as they do at Rocklands.

Men and horses are still used for the muster and they will usually manage to collect nearly all the cattle. In the past this was good enough as what remained in the paddock

Ian Braithwaite, one of the Stanbroke company vets.

Overseer Bob Buckley.

would turn up in the next muster. Now, however, it is not good enough and new techniques have been developed to ensure that not a single beast is left behind.

At Rocklands they muster the paddock with stockmen and horses in the usual way, but now the station Cessna is also used to flush out beasts that have been missed. After the muster the pilot flies Ray Jansen over every centimetre of the paddock in search of beasts that have still evaded the muster. If they can get them out of the paddock they do so, but if that is not possible these cattle have to be destroyed.

The aircraft is also used to check the many kilometres of fencing on the property. There is little point in doing a clean muster if other cattle can stray into the paddock through a broken fence.

Cattle have to have three consecutive clean tests before they can be declared disease free. With a minimum interval of sixty days between each test the tests cannot be completed in less than four months, and in practice they might take nearly a year. This means that the paddock will have to be mustered at least three times and on each occasion the cattle will have to be kept close to the yards until the results are known.

Until cattle have been declared free of disease there are stringent conditions on moving them off the property. They can be sent to the meatworks (where they perform their own tests); and they can be sent to other areas that have not yet been cleared of infection. But unless they have had two clean tests they cannot be sent to saleyards or to an uninfected area. The regulations are complicated (and often subject to different interpretations at local level), but they can never be ignored.

The result is that properties with infected cattle have to work them much more frequently than they did before. They must make sure fences are secure, especially if a neighbouring property is not as far advanced in eradication, and they need to subdivide their huge paddocks into more manageable sizes so that they can be mustered more easily. This is by no means as simple as it might seem. Fencing is expensive. In addition it will usually be necessary to sink several new bores to provide water to the smaller paddocks.

In spite of this, most cattlemen accept without question the need to eradicate these diseases. If that need results in changes to the way the place is run, then they are accepted as the price of economic and social survival. Nor are the changes without benefits. Cattle that are worked frequently become easier to manage and the commercial potential of the herd can be assessed more often. Nor was there ever much doubt that smaller paddocks would be easier to manage — it was simply that in itself that was rarely sufficient justification for the large capital investment required.

Stock woman.

Watching cattle.

Ray Jansen thinks Rocklands will be free of disease in about three years. Until then, their eradication programme will continue to dominate the way he runs the station and the success or failure of the programme will never be far from his mind. Indeed, as he is describing it a Cessna flies low overhead as it turns to land on the airstrip and a few minutes later a cheerful young man drops his bag in the spare bedroom and accepts a cup of tea in the kitchen. He is Ian Braithwaite, a Stanbroke company vet, and he has arrived to carry out the next batch of tests.

Born at Mackay in Queensland in 1959, he took his degree at the University of Queensland in Brisbane. Reluctant to go into suburban practice, he answered an advertisement that Stanbroke had circulated round the university. To his surprise he got the job and eighteen months later he is still delighted with it. He has a bachelor flat in Mount Isa but sees little of it as most of his time is spent on various Stanbroke properties.

The following morning he leaves the homestead at five o'clock to drive out to the Number 3 yards. They are 65 kilometres from the homestead and as he bounces along the dirt road the headlamps sweep across sleepy Shorthorns waiting for the day to begin.

It is nearly dawn by the time he gets to the outstation near the yards. He parks the ute and goes into the kitchen for breakfast, the light and chatter coming as a contrast to the still, dark night outside. He cuts some cold meat and bread and fills a mug of tea from the huge pot on the bench. He sits down at the long table and joins in the conversation, lively even at this hour.

Then there is a clatter of plates as men start to leave and outside, people climb into vehicles for the short drive to the yards as the sun edges over the horizon. By the time they reach the yards it is brighter, putting a yellow edging along the backs of the waiting cattle.

Men move quickly as they unload equipment from the vehicles, whilst others are already moving the first batch of cattle into the small yard at the head of the race, or greasing the sliding metal gates at either end of it. When they are ready, the first of the cattle are put into the race and the work begins.

An officer from the Northern Territory Department of Primacy Industry (DPI) smears glue on the back of a numbered piece of paper and sticks it firmly on to the back of the first beast so that the number is clearly visible. He moves from beast to beast, sticking numbers on each of them.

Meanwhile another DPI man has started behind him at the head of the line. Using a small scalpel, he makes an incision in the beast's tail to puncture the vein that runs along it. As the blood appears he collects it in a small glass

A road train waits to load cattle at Rocklands.

much bigger. The stockmen have also been drafting off weaners and calves, together with a few Africanders that should not be with this mob, so there are now smaller mobs in other yards. At 12.15 the last of the cattle are in the race and a few minutes later the job is finished. They have handled 904 beasts since 7 o'clock and they can now clear up and go back to the outstation for dinner.

During the afternoon the stockmen brand the calves and then turn the mob into a holding paddock where they will stay until the results of the tests are known.

Three days later the cattle are yarded again and passed through the race once more. The blood samples have been sent to Alice Springs to be tested for brucellosis and the numbers of positive samples have been sent back to the station. The cattle still have the numbers on their backs and they are identified as they go through the race. If the reading from the blood test is high, the beast will be destroyed, but if it is quite small the beast will be spayed instead.

The injection which Ian Braithwaite put into the tail of each beast was to test for tuberculosis. Now, seventy-two hours later, he examines the tail of each animal in the race to see if a small lump has formed. If it has, it is a sign that the beast might have tuberculosis. The beast is then destroyed and an autopsy carried out to determine if tuberculosis is present in the lungs. The reaction can be caused by organisms other than tuberculosis (and indeed usually is) and an autopsy is the only way of telling if tuberculosis is present or not.

There are usually very few positive blood samples or tuberculosis reactions, even in a mob being tested for the first time, and there should be even fewer on subsequent tests. Nevertheless, three clean consecutive tests will be required before they can be declared free of disease and if even one diseased animal turns up in the mob the testing procedure will have to go back to the beginning.

This is the second test and the mob is clean. The beasts can now be taken back to their paddock to look after themselves until their third, and hopefully final, test in three months time.

Ian Braithwaite drives back to the homestead to pack his things and to fly north to Augustus Downs, where he will work on another mob of cattle. He will be back at Rocklands in a few days time to inject more cattle that are now being mustered by Bob Buckley and his men.

Before he leaves he stops off at the horse yards to look at a mare recently bought by the daughter of a stockman. He tells her that it is in foal and if she is disappointed that she will not be able to ride it as much as she had hoped, she is more than cheered at the prospect of having a foal as well as a mare.

Then a stockman asks him to look at his dog, which has a wound on his leg. He does so and reassures the man that it should heal quickly but that he will look at it again when he comes back.

And that is about as close as Ian Braithwaite gets to being a suburban vet.

bottle which he holds underneath the tail. He then puts a stopper in the bottle and places it in a special tray that has holes numbered to correspond with those on the backs of the animals. He takes another empty bottle, looks for the beast with that number, and takes another sample.

Meanwhile Ian Braithwaite, looking incongruous in shorts, football jumper and a stockman's hat, has also started work at the end of the line. He injects the beast in the upper part of the tail and moves on to the next. There are several men taking blood samples but he is the only one giving injections and he has to work quickly.

When all the animals in the race have been attended to they are let out of the race into an empty pen. The race is then refilled with more beasts from the holding yard and they start all over again. It is skilled, repetitive work. As it goes on the sun gets higher and hotter and dust hangs over the cattle as they are moved. The men have a short break in the middle of the morning and then start again. The mob of treated cattle grows, but the mob waiting to go through the race seems no smaller now than it did at the start.

Then, almost suddenly, this changes and the waiting mob is now noticeably smaller whilst that in the collecting yard is

ELGIN DOWNS

As the men drive out to the horse yards a kilometre from the homestead they are grateful for the clouds in the sky. Although Elgin Downs is 160 kilometres north of the Tropic of Capricorn in Queensland the clouds have put a slight chill in the early morning air. It will not last long. Within an hour or so the sun will have broken through and the temperature will be climbing steadily, but until then they can enjoy the coolness as they start the day.

Although this country has been settled for more than a hundred years, the four properties that make up Elgin Downs came under common ownership only about twenty years ago. But the story of how that came about starts more than eighty years ago, and it starts not in Australia but in Texas, in country fairly similar to this.

At the turn of the century one of the biggest cattle companies in America, King Ranch, was facing a dilemma brought about by the harshness of its Texas land and the increasing demand for good quality beef on the American market. They had always run the hardier breeds of cattle but the coarse beef they produced was losing its appeal on the market. Knowing that pure bred English cattle produced beef of better quality, they imported them in sufficient quantity to stock the whole of their land.

It was a disastrous move. They soon found that these aristocratic beasts could not stand the heat of the midday sun and that they lacked the immunity to disease enjoyed by the hardier breeds. Although producers of prime beef in their home country, they could barely survive in Texas.

Some of the permanent water at Elgin Downs.

N

← To Charters Towers

Ti Tree

Sutton

Three Mile

Spectacle

Bells Lagoon

DISNEY

AVON DOWNS

Durana

Bottom Mistake

Sandy

Georges Plain

Holding

Yarmina

Top Mistake

Mailbox

Gidyea

Diamond Creek

Belyando R.

Bottom Nibbin

New Pdk.

Bottom Mistake

Mistake Ck.

Pocket

NOT PART OF PROPERTY

Top Nibbin

Six Mile

North Plain

Mountain

ELGIN DOWNS

NEW TWIN HILLS

Rossmore

HOMESTEAD

South Plain

Back Washpool

Highway Pdk.

To Clermont →

Fox Ck.

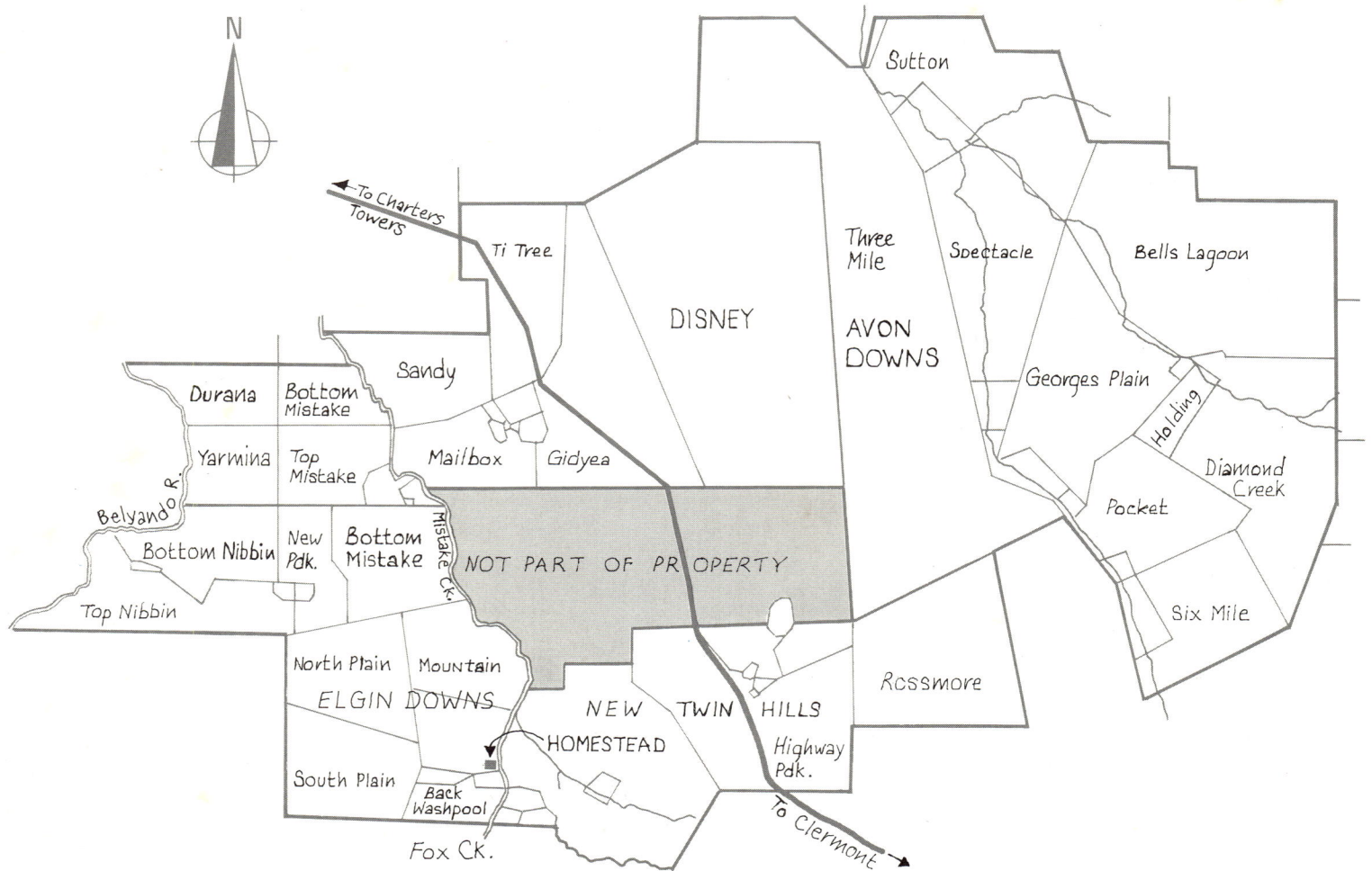

The solution was obvious to everybody: what they needed was an animal that could produce good meat and which could thrive in that hot and arid country. The difficulty was that no such animal existed. Different breeds met some requirements more than others, but no single breed met them all. And so the men of King Ranch set out to produce one.

It was a long and highly skilled job. Different breeds were crossed and the progeny studied for improvements. In the end, after much trial and error, they found the answer by crossing British Shorthorns with tropical Brahmans. The result was a young bull which successfully combined the heat resistance of the Brahman with the quality and temperament of the Shorthorn. He was called Monkey, and he was the foundation sire of a new breed which they named after a creek that ran through their land — Santa Gertrudis.

In the 1930s a Santa Gertrudis bull was sent as a gift from King Ranch to a scientist who was carrying out similar experiments in cross-breeding on a cattle station in Queensland. The progeny of this bull were seen by Sir Rupert Clark and during a visit to America in 1951 he suggested to Bob Kleberg, then head of King Ranch, that they might form a partnership to establish a Santa Gertrudis stud in Australia. The idea appealed to Kleberg and the partnership was formed the following year, although by then they had decided they would run a commercial herd as well as a stud.

The first shipment of 45 bulls arrived in Brisbane the following year and was sent to the new headquarters of King Ranch Australia at Risdon, near Warwick in Queensland. Later that year another 230 beasts were landed at Melbourne and they too were sent to Risdon, a rail journey of more than 1,600 kilometres. In 1956, however, the Commonwealth government put an embargo on the importation of live cattle and the new venture had to rely entirely on the 326 beasts that had already arrived.

Meanwhile, King Ranch had been looking for properties in Australia which would be suitable for this new breed, and they found them in the heavily timbered country north of Clermont, about 300 kilometres inland from Mackay. In 1952 they bought two adjacent properties, New Twin Hills and Elgin Downs, which between them covered 663 square kilometres of gidgee and brigalow scrub.

In 1955 they appointed a well-known Queensland cattle man, Jack Cooper, as manager. Three years later they bought a 953-square-kilometre property, Avon Downs, and Jack Cooper's son John was brought in to manage it.

This was not nepotism, for John Cooper had already established himself in the Queensland cattle business. Born in Brisbane in 1926, he was educated at Nudgee College and when he left in 1941 he went to Marodian as a jackeroo for three years. He then worked for his father, who at that time was manager of Sanders Station near Dingo. In 1955 he

Branding a calf.

John Cooper, manager of Elgin Downs, boils the midday billy.

Drafting a mob of Santa Gertrudis.

An old outstation on Avon Downs.

married and went as assistant manager to Barkly Downs near Camooweal, where his wife Shirley was one of only two white women for hundreds of kilometres. By the time he took over as manager of Avon Downs he was thirty-three years old and had spent his working life with cattle.

In 1964 King Ranch completed their purchase when they bought the property of Disney, a 585-square-kilometre property that linked Avon Downs with Elgin Downs on the northern side, as New Twin Hills did to the south. They now had a property of half a million acres, although in the middle of it a property of 51,000 acres, Old Twin Hills, remains privately owned.

King Ranch had also been buying properties elsewhere and in 1965 they appointed Jack Cooper pastoral superintendent of Queensland and John Cooper took over as manager of the combined properties and moved on to Elgin Downs. He has been there ever since.

Although this land had been running cattle for years, it was virtually unimproved when King Ranch bought it. As such, it had been capable of supporting cattle on a fairly modest scale, and with the huge area involved that was all anybody had needed of it. Until now.

Whilst the northern part of Avon Downs was open coolabah country which needed little improvement, most of the rest consisted of heavy gidgee and brigalow scrub. This was so prolific as to make it almost a jungle and it was this

that had kept the stocking rate low. If it could be cleared, even in part, the improvement would be dramatic indeed.

Until now nobody had attempted to clear this kind of scrub on such a large scale. The costs were far beyond the reach of individual owners and there was no certainty that the huge investment would ever be repaid. Regrowth of the scrub after it had been cleared was certain to be a problem and the costs of keeping it cleared might go on for years. King Ranch, however, had two advantages. One was that they had experience of clearing similar country in Texas, and the other was that they could afford to do it.

On the face of it the job is simple enough. A huge D8 tractor positions itself on the clear land at the edge of the scrub, whilst another goes into the scrub and takes up a position level with the first and about 60 metres away. A heavy chain is run between them and a seven-tonne ball is fastened to the middle of it. When everything is ready the tractors move forward at about three-quarter power and the ball and chain clear the scrub as they go. The thin spindly trees crash to the ground as the two tractors bore on like primeval monsters in a Hollywood movie.

The jumble of fallen timber is metres high but after it has been left to dry it burns readily and the vastness of the country can be seen for the first time. There is, of course, very little pasture on it for what little there was has been removed in the clearing of the scrub.

The homestead at Elgin Downs.

The next stage is to turn the open country into good grazing land. This is done by sowing huge quantities of buffel grass from the air. Buffel grass will grow on poor soil in almost any conditions and yet provide good and reliable feed. If the new breed of Santa Gertrudis cattle was one of the keys to successful cattle-raising in tropical Australia, then buffel grass was the key to providing sufficient feed on cleared land to make the stocking rate economic.

It takes between two and three years to turn virgin scrub into good grazing land. Even then, the paddock of knee-high grass will look rough and uninviting, but for beef cattle the paddock will be as good as a Hilton. It will have cost about $12 an acre. It doesn't sound much until you realise how many acres there are. There are about eighty paddocks at Elgin Downs and they range from 150 acres to 40,000. A paddock of 15,000 acres costs about $180,000 to clear and it will be out of use until the buffel grass has put on enough growth to support the cattle.

Faced with costs like this even King Ranch cannot afford to start at one end and work right through. So far they have cleared about 150,000 acres and the work goes on each year.

It is, like mining, a form of resource development, and the costs are not much different. What is different is that at the end of it you can watch Santa Gertrudis cattle, as they in turn watch you. Today, the Elgin Downs complex runs about 27,000 head of cattle and about 700 horses, but it would not

have been possible without Santa Gertrudis, buffel grass, and a few million dollars.

By the time the men have saddled up and mounted, the clouds have left the sky and the coolness has gone. It is going to be a hot day, but there is nothing unusual about that. The horses look superb as they stand there patiently flicking their tails and occasionally shaking their heads as they wait for their day to begin.

They are Australian quarter-horses bred from imported American horses crossed with Australian stock horses. They have an instinct for cattle work and often horse and rider will decide on the same course of action at the same time without any communication between them. These horses have been bred at Elgin Downs and, given a free choice, the men would still use nothing else. If part of your working day is spent putting a horse at full gallop in heavy scrub, you tend to have firm views about the sort of horse you want to be on.

As the men mount up, John Cooper gives last minute instructions about the day's mustering. He is a perfectionist of the old school and he expects his men to be the same. There is only one way of doing a job and that is the right way — first time. Those who disagree move on fairly quickly, those who stay are often with him for years.

The men ride off, the horses' hooves putting up rhythmic spurts of dust so that soon there is a moving cloud of it just above the ground. The men sit easily in the saddle, moving automatically as they chat about the movie they saw the previous evening. It is a fortnightly event and brings in men from the other two homesteads on distant parts of the property.

John Cooper watches them go, then adjusts his high-crowned straw hat, calls his two Labradors on to the back of the ute, and drives back to the homestead. It is a modern air-conditioned house built on stilts so that the ground floor is open space, with the station office tucked into one corner. A staircase leads up to the house above and opens out on to a big screened verandah. It is dotted with pot plants and outdoor furniture and is as much a room as those inside.

Outside, a huge generator produces electricity for this and other houses, for there is no mains power here. The generator starts at 5 a.m. and goes off at 11 p.m., although on most days it goes off in the afternoon as well.

In the big lounge of the homestead a colour television stands against one wall but in this isolated spot some 135 kilometres from Clermont there are no transmissions for it to receive. Instead it shows movies from video cassettes which are brought out once a week and shared with others on the property who have similar equipment. If it is a long film, they put it on early so that it has finished before the power goes off at eleven. After a long day, this is late enough for most people.

John Cooper goes through to the kitchen to collect a packet of sandwiches. Alongside the kitchen is the men's dining room, where he usually starts his day by having breakfast with them. It is empty now, for it is past eight o'clock and the day is well advanced. But this is one of the days when the power will stay on all afternoon and the kitchen is busy as Shirley and the cook get ready for a day's baking. It will take most of the day, but with the men out mustering they will have an uninterrupted run.

John leaves them and goes downstairs to the ute. He is going to spend the day driving round part of the property. He tries to see all of it every two weeks and in order to do this he has to be out nearly every other day. Today he is going to part of Avon Downs.

The nearest paddock is 32 kilometres away and the road is quite good, but once inside the paddock only he would know there was a road at all. This part of Avon is still uncleared brigalow and the track is barely visible in the undergrowth. The ute rears and plunges over the uneven ground as he drives slowly along, looking for cattle.

Although there are several hundred head in this paddock it is ten minutes before he sees any. Then a dozen bullocks look at him quietly from the scrub and others nearby lift their heads in curiosity. And as he looks at them, John Cooper starts talking to them, softly and automatically repeating phrases that stockmen use when they are working them on horseback. It is an old habit that dies hard, even though the engine of the ute now drowns out the sound.

As he drives slowly through the paddock he analyses everything he sees. When he looks at cattle he can tell

Santa Gertrudis cattle. Bred by King Ranch in Texas, Santa Gertrudis were first imported into Australia in the early 1950s.

whether they are in prime condition or recovering from a feed deficiency brought about by lack of rain, or falling off for the same reason. By looking at the feed in the paddock he knows how long it will support the cattle and he can tell from any new growth when it last rained. He is like a businessman reading a balance sheet, for his is written here.

He stops at all the dams to see how much water they are holding and to make sure it is being pumped into the long trough nearby. At one of them he sees a group of wild pigs a few hundred metres away and lifts his rifle from the bracket behind his seat. He leans across the bonnet and fires. He misses, as he thought he would, but it was worth a try.

By the middle of the day he has left the scrub behind and is running across the vast openness of Georges Plain. It makes up much of the northern part of Avon Downs and is prime country which never needed clearing. In the distance a row of trees marks the fence line but it is far away and even the trees are no more than a smudgy green line shimmering in the heat.

A little further on there are more trees and soon he is in wooded country — but these are coolabah trees, not scrub. He stops the ute beside a soak and fills a blackened billy. He lights a fire, sets the billy, and then walks over to inspect an old timber well, long since dry. Inside a python resents the intrusion and starts to move down an inclined plank to more peaceful regions below. Stretched along the plank, it is fully three metres long and beautifully patterned.

John goes back to the fire and throws some tea into the boiling water in the billy. He then whirls it round his head to make the leaves settle and pours himself a mugfull. After his meal he lays on his back on the ground, puts his hat over his eyes, and goes to sleep for ten minutes. He is only a metre from the well and a python that could be anywhere.

It is five o'clock by the time he gets back to Elgin Downs. During the day he has driven about 400 kilometres and all but 80 of them have been on the property. But he has not finished yet. He drives past the homestead and goes out to the cattle yards to see if the men have brought in the cattle.

The yards are still empty but he hears the sound of approaching cattle. Soon he can see a large mob of cows and calves being walked slowly towards the yards. A single rider leads the way, whilst others are spaced out behind them. Calves skitter around the edges of the mob, sometimes almost hidden in the cloud of dust.

As they approach, a man rides ahead to open the gate into the yard. The others split the mob into small groups and then allow each group to pass easily through the gate. If they let the whole mob through the gate at once some would almost certainly be injured on the posts from the weight of those behind. When all the cattle are in the yards they close the gate and the assistant manager rides over to talk to the boss as the others turn for home.

John Cooper and the men are back at the yards at 6.45 next morning preparing for a day's drafting and branding. In the middle of the yards is a race. Sliding metal gates are greased so that they will push across easily, other men

Yarding a mob at the end of a muster.

prepare the dip which makes a long bath at the far end of the race, whilst others unload more equipment from the utes.

When everything is ready, the men move the cows and calves into another yard at the head of the race. A little way along it, where several gates lead from the race to different yards, John Cooper is standing with a couple of men. What they do now depends entirely on which animal comes along the race.

If it is a cow, she passes the gates and stops at the dip. Men there slap her rump to make her plunge in and she swims through it before climbing up the ramp into a small circular yard. The dip will protect her against ticks and later, when she has been joined by more cows, she will be sprayed against buffalo fly before being released into the bigger yard beyond.

If it is a calf, a man at the head of the race checks to see if it has been branded. If it has, it is now old enough to be weaned from its mother and a gate is opened to direct it into one of the yards. If it has not been branded then it has been born since the paddock was last mustered, and another gate is opened for it to pass into a third yard.

A few hours later all the animals have passed along the race and are in one of three yards: the cows are in the yard beyond the dip, the weaners are in a yard half way along the race; and the unbranded calves are in another yard on the opposite side.

These cleanskins are now taken back to the head of the race and passed along it again one at a time. It is the most important part of the work, for John Cooper, using an

152

awning beside another race. A gas bottle is connected to a plate with burners which will heat the branding irons, whilst men bring more equipment from the utes. When all is ready, the calves which will become stud bulls are led into the race and held behind a sliding gate. On the other side of it is a metal frame which is hinged at the bottom and shaped to hold the calf. The gate is slid open and the first calf steps into the frame. The frame is immediately closed around it and two men pull it sideways and downwards until the calf is lying on its side on the ground.

One man holds a hind leg to stop the calf moving and John Cooper immediately applies the station brand to its side. As he does so, another man calls out the number for this calf and starts to pass him the sequence of numbered irons so that he can brand the number below the station brand. At the same time another man is tattooing the same number inside the calf's left ear, and then tattoos the letters ELG inside the other. Meanwhile, another man is giving the calf an injection below its shoulder to protect it from tetanus and other diseases.

It is highly efficient teamwork and the whole operation takes no more than thirty seconds. Then the cradle is opened and the calf bounds to its feet and runs out into the yard. The cradle is swung upright again as each man prepares for the next calf. It goes on for hours and the men sweat and grunt as they throw the cradle to the ground. But the efficiency remains and the brand on the last calf is as straight and evenly spaced as it was on the first.

When all the calves have been branded they are again passed along the main race so that they go through the dip before joining the cows in the far yard. They will be left together until each cow has found her own calf and then they will be taken back to their paddock.

The weaners will remain in the yard on good feed for three days. They will then be taken to their new paddock and kept in a mob for another couple of days to get them used to being worked by men and horses. They will then be drafted again, with the bulls, heifers and steers going into different paddocks.

Meanwhile, it is the end of a long, hard day and the men quickly pack the gear into the utes and drive home. Tomorrow they will be mustering more cattle and later in the week they will be doing the same work with them in the yards. Some paddocks are so big that they take weeks to muster and some are so far away that they use other yards that are closer than these. It is all part of the routine of a cattle station, and after spending all day with cattle you don't need to stand and look at them when you have finished.

But John Cooper does, because he wants to. He leans against the rails and looks at the weaners pawing the dust in the yard and he talks to them softly, gentling them down for the first night they will spend without their mothers. Appealing in their immaturity, they look back at him with big eyes, their ears twitching slightly to catch the sound of his voice — gentle and reassuring in the fading light.

experience developed over the years, has to assess each one. In doing so, he is largely determining the future success of Elgin Downs.

The best male calves will be retained as bulls. They will grow to about 900 kilograms and in due course they will either be sold to other breeders or retained for showing. They sell about 400 bulls a year at Elgin Downs and they have never had a buyer on the property who has not bought at least one animal.

The other male calves will be castrated to grow up as steers and then run as bullocks in the commercial herd. They will be grown for meat and that is why they are turned into steers. As such, the meat they produce will be of high quality, whereas the meat produced by bulls is tougher and almost worthless.

The female calves are also selected in the same way. The best of them will go into the stud for breeding, whilst the others will go into the commercial herd as meat producers.

As each calf comes to a halt before him, John Cooper has to decide its category. In some cases it is obvious and he calls out so that a man can open the gate for the appropriate yard. Others are not so obvious and he takes several minutes to make up his mind, often discussing the animal with his assistant manager before making a decision. It is particularly important with a male calf for once having turned it into a steer there is not much anybody can do about it.

When all the calves have been classed the men prepare to brand them and set up the equipment under a thatched

THE SCHOOL OF THE AIR

The high, piping voice is remarkably clear as it comes through the loudspeaker in the studio.

'Good morning, Miss Freeman, girls and boys.' Miss Freeman sits at her microphone and looks unseeingly across the studio, her mind tuned, like the radio, to more distant parts.

'Good morning, Jimmy. How's your dog today? Over.'

'He's a lot better, thank you, Miss Freeman. Over.'

'That's good. I'm sure he will be all right now. Everybody please open the story book at the first page. Jimmy, as it is about a dog perhaps you should be the first to read. Over.'

'"Woof, woof," said Spot, and he ran and jumped over the wall into the garden...'

Beyond the glass wall of the studio tourists sit in the visitors' room and listen as Jimmy reads the first page of the story. They are silent, as they would be in any classroom, careful not to interrupt and break his concentration. For unusual though this school is, they know it is a *real* school and that the lesson being conducted here is no less serious than those taking place in conventional schools throughout the country. The difference is that this classroom covers more than a million square kilometres of outback Australia.

The first School of the Air in Australia started here in Alice Springs in 1951. Until then, children in the outback were educated by correspondence courses sent

A teacher conducting a class in the School of the Air studio at Alice Springs.

to them by State Education Departments. It was a good system (and still is), but there were unavoidable weaknesses. As it had to rely entirely on the written word, children with reading difficulties were at an immediate disadvantage. Indeed, if a child was the only one on the property it could be very difficult, without other children for comparison, to even recognise the problem.

It was a South Australian school inspector, Adelaide Miethke, who on a visit to Alice Springs in 1946 first saw the possibility of using radio to add an extra dimension to outback education. When she met the director of the Royal Flying Doctor base, he told her that a nursing sister had recently used their radio network to give a very successful talk on child care to mothers on distant properties.

Adelaide Miethke remembered two children she had seen on a property during her recent journey. Unused to seeing visitors, they were too shy to talk to her and she had realised then how different their lives were.

'If you could talk to mothers all at once, couldn't you talk to children?' she asked.

'Yes, I suppose so.'

She thought of children and stories and things from the outside world — and she thought of the School of the Air.

Today, there are twelve Schools of the Air operating throughout the remoter parts of Australia. They bring primary education to isolated children to supplement the correspondence courses and provide contact with other children and a 'real life' teacher. All but two, at Alice Springs and Katherine, still use the facilities and equipment of the Royal Flying Doctor Service. These two are also unusual in that they issue correspondence courses and make visits to their pupils, tasks which elsewhere are carried out by special units within the States' Education Departments.

For Fred Hockey, principal of the Alice Springs School of the Air, it is the most exciting job he has ever had. 'Parents on these properties are anxious to give their children a good education and they are prepared to work at it. On our part, we feel for the kids in isolated areas and we don't want them to be disadvantaged because of distance.'

Teachers are drawn from State schools and must have had five years teaching experience before they will be considered for the School of the Air. If they are selected they serve on probation to make sure they are suitable. Even so, there is no shortage of applicants and there is usually a lengthy waiting list of teachers wishing to join.

Teaching methods are, of course, quite different. The teacher has to operate a sophisticated studio console

Although he lives on an isolated property in the outback, this boy has a class with the School of the Air every day.

whilst conducting the lesson, bringing in music or other recordings as the lesson requires. With no blackboard, the teacher has to rely entirely on the voice as the means of teaching: he cannot draw diagrams, write letters or point to objects. And because the lessons are often heard by others on the property who might have the radio on standby, a teacher cannot easily reprimand a child as this could be a public censure of dreadful importance.

Using a transceiver supplied by the School, every child joins in a group lesson lasting between twenty and thirty minutes each weekday, and those aged between five and twelve have a private lesson of ten minutes each week. The School expects the child to work under the guidance of a supervisor, who will also teach the correspondence course.

The supervisor is often the child's mother, but a full-time governess might be employed by the station if there are a number of children on the property. Governesses are often girls who have recently completed their own high school education and who therefore have no training as teachers, although an increasing number of trained teachers are now taking up this work.

Recently the School issued video cassette players to properties that use the School of the Air. Educational videos are supplied and are then discussed over the radio during class sessions. This extension of the School's service has again added a new dimension to education in the outback. It is greatly appreciated (especially by those too remote to receive off-air television) and children are now able to see, often for the first time, something of the 'outside' world.

The teachers at the Alice Springs School of the Air try to visit each child at least twice a year and this can involve considerable travelling. Although each visit might last only a day or two, they are of very great importance. The child is able to talk freely to the teacher (as well as having the unusual thrill of their 'own' visitor), whilst the teacher can make a professional evaluation of progress and help with difficulties that might be beyond the skill of the supervisor.

Surprisingly perhaps, the 130 children taught by the School of the Air at Alice Springs, and those taught by other Schools elsewhere, receive an education which in many ways is better than that available in conventional schools. Teachers are able to develop a unique relationship with each pupil and specific difficulties are likely to receive much more attention than they would in a city school.

The School at Alice Springs even provides accommodation at the School so that supervisors can bring to the school any child who might need special help. The staff can then spend several days testing the child, working out remedial teaching programmes, and advising the supervisor on their use.

Fred Hockey, Principal of the School of the Air at Alice Springs.

On the other hand, children taught by School of the Air, or by correspondence alone, can lack competitiveness or have difficulty relating to other children, so that for some the eventual transition to a boarding school for their secondary education can be quite traumatic.

The School tries to overcome this by organising a 'Get-Together Week' which is held in Alice Springs at the end of September. It is the highlight of the year and children travel hundred of kilometres to attend. They meet their teachers and, perhaps for the only time in the year, the other children in their class. There are camps, film shows, excursions, picnics and, at the end of the week, a breaking-up party.

Meanwhile, back in the studio Miss Freeman is bringing the lesson to a close.

'That was very good, Gillian. Thank you. And that's all for today, boys and girls. Same time tomorrow. Goodbye.'

And from the loudspeaker a dozen small voices pipe back from a classroom that is four times the size of England, 'Goodbye Miss Freeman.'

In tiny rooms in distant homesteads, or in improvised schools beyond the cattle yards, children put down their microphones, switch off the transceivers, and look forward to tomorrow.

GOGO

There was nothing very unusual about the weather at Gogo on Thursday, 17 March 1983. In this part of West Kimberley, as in other parts of the Top End, the Wet usually lasts from about November to March, but it varies from year to year. It is an aptly named season for most of the year's rain falls during that time and heavy storms are common. Unable to move along boggy roads or across sodden paddocks, most station hands are stood down and those that remain spend their time doing maintenance work in anticipation of the coming dry season, for there will be no time to do it then.

It is an unpleasant time of year. The temperature hovers in the 40s for days on end and even the storms bring little relief. The heat and humidity produce insects by the thousand to torment people and cattle alike. This soggy, dripping world is rarely seen by tourists, who sensibly wait for the dry season which follows. But it is only too familiar to those who live on the stations.

There is no dramatic end to the Wet. The storms and the rain simply become less frequent and as the days become dry and sunny people realise that the Wet is at last over for another year. Sometimes it deceives them. Clouds might build up again until they cover the sky and they know that there are still a few more storms to come before they can finally say that it is over.

It had been like that at Gogo in early March. It was still too early for the Wet to be over but the days were bright and dry and the Wet was clearly coming to an end. There would be a few more storms perhaps, and then the stockmen would

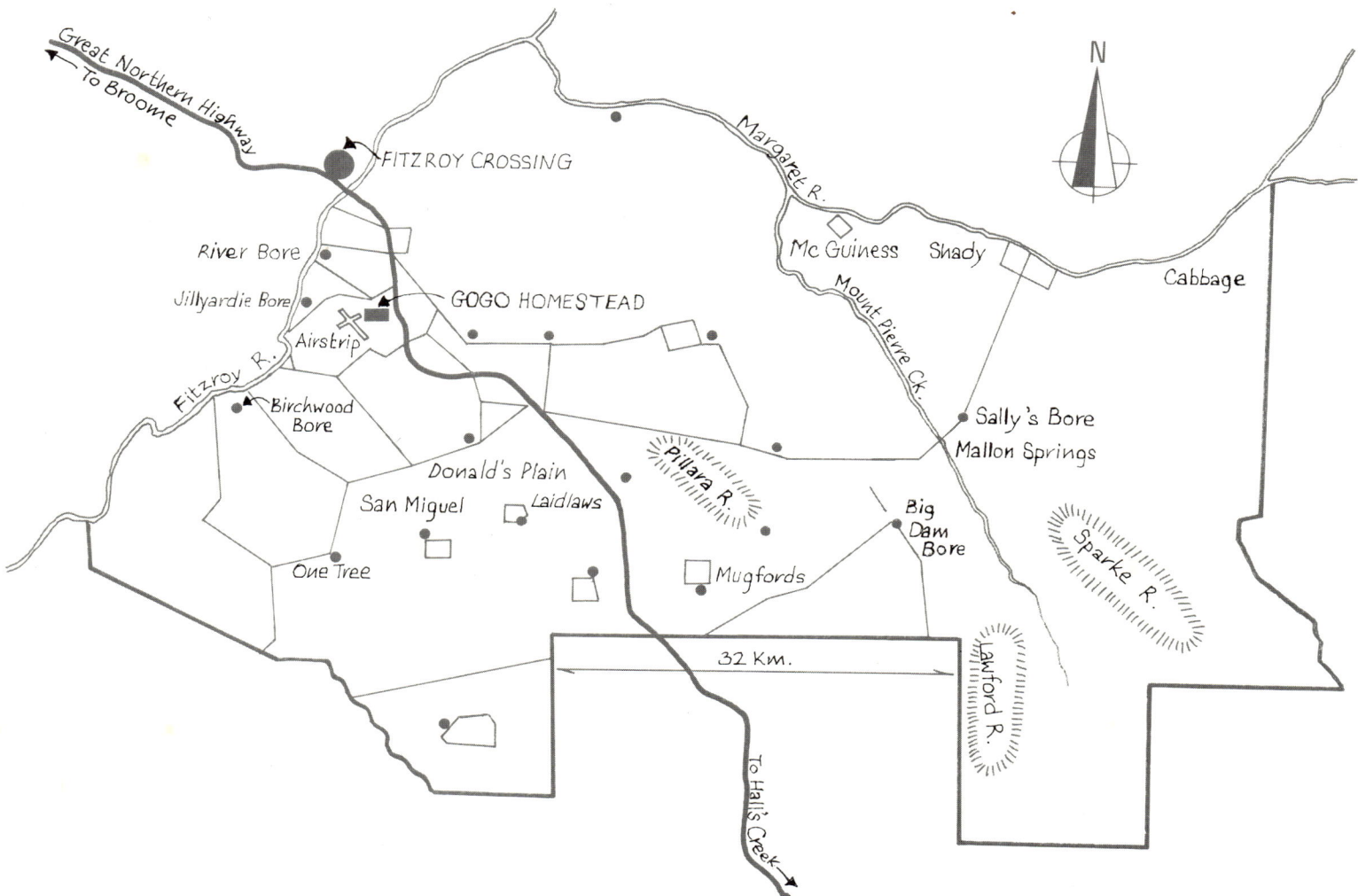

return to the station and everybody would gear up for the busy work of the winter months. This is the only time stock can be sent to the meatworks and the station has to earn the whole of its annual income during these months.

On 17 March, then, nobody was surprised at the weather. Clouds were heavy overhead, lightning flashed for most of the day and from time to time thunder rolled across the homestead and out into the paddocks beyond. There was some rain but it was intermittent and not particularly heavy. The creeks and watercourses were running swiftly from earlier rain but were no higher than expected at this time of year. Until the end of the day.

Then Len Hill, the manager of Gogo, noticed that they were rising. At first it seemed simply the result of the day's rain, but as they continued to rise it became obvious that it was more than that. The rain that had fallen that day, useful though it was, had not been enough to have that effect, and certainly not for long. As he made his way round the paddocks close to the station area he saw that many of the creeks were over their banks and the water was starting to spread across the flat ground that separated them.

He knew now that the water must be coming down the two rivers that bordered the property: the Margaret River on the northern boundary, which joins the Fitzroy River to make the boundary on the western side. Their headwaters were far away, high in the hills of remote Kimberley country where there was nobody to monitor their flow. If there had been heavy storm rain in those hills, which now seemed likely, nobody would know until the rivers started to rise further down. And now they were.

Even that was not very unusual. In most Wet seasons these watercourses go over their banks and water covers the flat ground nearby. Often it would happen in the farther reaches of the property and they would not even know of it until the Wet was over and, able to move freely once again, men would see signs of minor flooding.

Occasionally the flooding would be more serious. In 1956 much of the country had been covered with water and the damage had been considerable. It was the worst flood they had ever had and somebody had marked the level of it on the wall of an old station building. It seemed unbelievably high and people sometimes looked at it and wondered what it must have been like.

Unable to move very far, for the black soil was thick and glutinous, Len Hill watched the water rise. It had not yet reached the top of the bank where he was standing, but the water was running fast and turbulent. It was carrying down a lot of debris and the surface was covered with swirling logs which frothed the water as they were swept along. Then, as the gloomy light faded to even gloomier darkness, he saw something else.

In the middle of the water was a cow. Her head was just clear of the water as she was carried along with the flow, unable to reach the slower water at the edge of the creek. Then behind her came another, and then another. And Len Hill knew that the rushing water had swept them from their

The debris caught in the tree shows the height of the water which swept across Gogo in March 1983.

paddock and carried them away, some to drown as they were swept along, others by some circumstance of chance to reach safe ground further down. And he knew for certain now that Gogo was flooding.

By the following day most of the paddocks were under water. The homestead area, built on higher ground, was still dry but it was surrounded by water and travel was impossible. Between the huge expanses of water, which was still flowing quickly even though it now covered a vast area, cattle had gathered on the higher ground which stood like islands between the lakes. Through binoculars Len Hill could see a lone cow standing on a small hummock which got smaller as he watched. The water rose, lifted the cow from her feet, and carried her away to God knows where.

During the afternoon the water rose 30 centimetres every fifteen minutes and had long since passed the level of the 1956 flood. 'We simply looked at in awe,' says Len Hill. 'There was nothing else we could do.'

It started to go down the following day and by Sunday the water had returned to the confinement of creeks and rivers and they were able to move about again. A man went to the old station building and marked a line across the wall to show how high the water had been. He painted the date at the end of the line and then measured its height above the line that had been painted in 1956. It was two metres higher.

The station area, Gogo.

Max Ley, station pilot, mustering cattle with a Piper Super Cub.

Alexander Forrest was the first European to see the strange and beautiful land of the Kimberleys when he explored the country between the deGrey River, near what is now Port Hedland, and the Overland Telegraph Line south of Katherine in the Northern Territory in 1879. His account of the journey described the Kimberleys in glowing terms. It was well watered, it carried abundant growth, and it was available to anybody who could stock it.

Early in 1881, on his property Thylungra in southern Queensland, an Irishman called Patrick Durack read Forrest's account and thought it would be an ideal place to start a new station. He had pioneered that part of Queensland in 1867 and now, with his property well established, some of the excitement was missing. His wife was less certain, however, and suggested that when he was next in Goulburn he might talk it over with his old friend, Solomon Emanuel.

The Emanuels had settled in Goulburn in New South Wales in the 1840s and soon became involved in a wide range of businesses. When Patrick Durack had bought his first land there it had been on a mortgage from old Samuel Emanuel and later, when Patrick Durack had headed north for Queensland, he did so on finance provided by Samuel's son, Solomon.

Patrick Durack soon found a compelling reason to visit Goulburn, and when he talked to Solomon he found him almost as enthusiastic as himself. Solomon had two sons, Sydney and Isadore, and he might be interested in taking up land in the Kimberleys on their behalf. He suggested that if Patrick Durack and his brothers were to organise a private journey of exploration, then he would share in financing it.

After a quick visit to Perth to talk to Forrest himself, Patrick returned with even more enthusiasm and immediately started to plan the expedition. It would be led by his brother, Michael 'Stumpy' Durack, who was an expert bushman, and Sydney Emanuel would be part of the team. The cost was estimated at £4,000 and this would be shared equally between Patrick Durack and Solomon Emanuel.

Meanwhile, 'map speculators' in the eastern States were already claiming huge areas of this new land in the Kimberleys without even seeing it. Solomon, concerned that the best land might have been taken up by the time the expedition arrived, suggested they did the same. They claimed a holding of 1½ million acres on the Fitzroy River in West Kimberley and a similar amount on either side of the Ord River in East Kimberley in the hope that the land might prove to be as good as it seemed from the sketchy maps.

Durack's expedition returned to Sydney in 1883, having inspected most of the land they had claimed. It was every bit as good as they had hoped and all that remained was to decide who would settle where. In the end it was agreed that the Emanuels would take the land on the Fitzroy River, whilst the Duracks would settle on the Ord River to the east.

Meanwhile, other adventurous pioneers were taking mobs of cattle overland to this promising new area and most of the good land was soon occupied. By 1887 Isadore Emanuel had several properties in the Fitzroy area beside Gogo and five

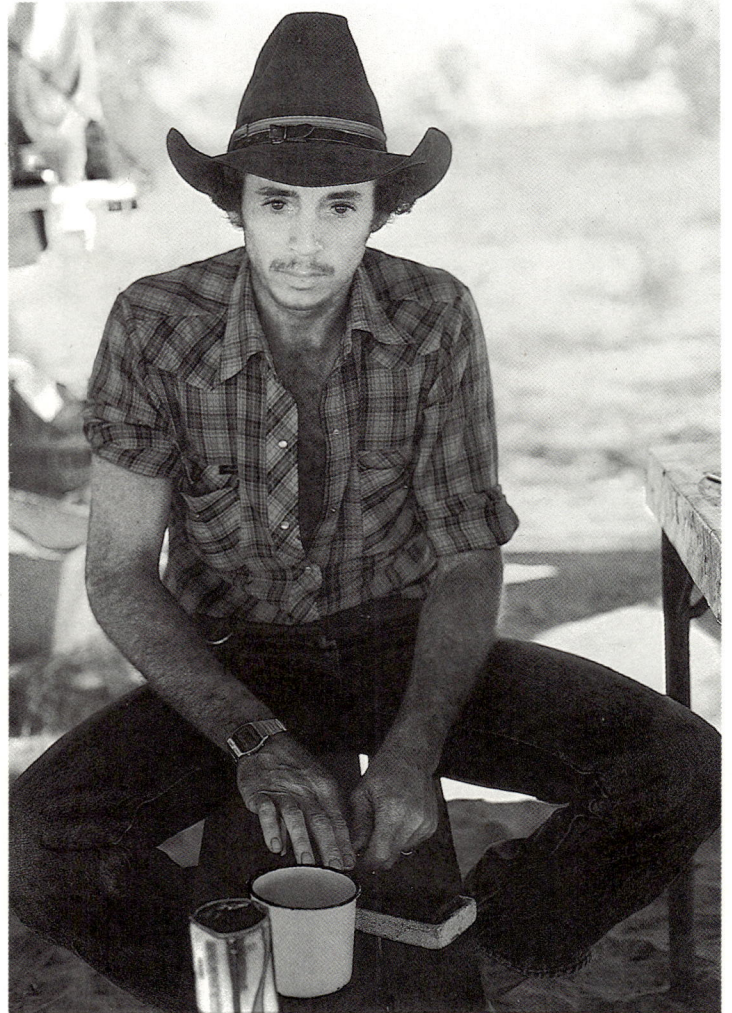

Michael Clinch, head stockman.

years later the Emanuels were the largest single leaseholders in the Kimberleys, although Isadore said that so far they had not made a penny out of their £50,000 investment.

In 1890 the Duracks were ruined, partly as a result of Patrick's speculation on the new goldfield at Halls Creek but more particularly because he was caught up in the widespread financial collapse in the eastern States. In the west, however, Isadore Emanuel was able to consolidate his holdings and when gold was discovered at Kalgoorlie he was in a good position to take advantage of this new market.

But with prosperity came criticism, as many people suggested that he might be taking too much advantage. There were many complaints about the high price of meat in Western Australia and accusations that it was because the Emanuels had too much control of the market. A Royal Commission was set up and Isadore gave evidence before it, but it was the new Labor State government, elected in 1911, that made the dramatic changes by passing legislation prohibiting individuals to own more than one million acres.

Isadore was obliged to sell many of his properties but he was unable to find buyers for the cattle stations and he had no choice but to keep them. Incensed by the new government, however, he moved to England in 1912 and continued to run the properties from there until his death in 1954.

Len Hill, manager of Gogo.

His son Sydney then took over. He had been born in Perth in 1903 and was only seven when the family moved to England. Although he now made frequent visits to the Kimberleys he continued to run the properties from England and for most of the time was a very absent landlord.

It was his son, Tim Emanuel, who brought the management of the properties back to Australia. He was born in 1936 and after doing two years' national service in the army he moved to Perth to control an empire that now covered four million acres.

Gogo, one of the four properties that make up that empire, consists of 913,000 acres of good Kimberley country which straddles the Great Northern Highway east of Fitzroy Crossing. In the south of the property the country is mostly red and black soil and is fairly flat. Further north it tends to be hillier and ranges there and to the west stand high above the surrounding country, dominating it with long escarpments that glow red in the afternoon sun. The country is rougher between these ranges, but it is still good cattle country.

Mitchell and Flinders grass grow well and there are areas of buffel grass which was introduced in the 1950s and which has since spread on its own. Spear grass is also abundant. It grows to about 2 m and stands thick and high, the dark tips shifting slightly in the breeze. There are trees around the watercourses and occasional patches of scrub, but most of the country is open. It has never been cleared and the pasture does not need rejuvenating provided it is not overstocked.

Gogo is divided into eight paddocks and most of them are huge. The biggest covers 409,000 acres, or 1,657 square kilometres, and is really open range country rather than a paddock, especially as much of the boundary remains unfenced. The property has a rainfall of 18 inches a year, most of which falls in the Wet, and the country is well watered by the Fitzroy and Margaret Rivers. Twenty-five bores provide water in the winter when the creeks that lattice the property have run dry.

The Wet produces lush and prolific growth and when it ends the country is green and abundant. The growth dries out during the winter and the colour of the landscape slowly changes from green to brown, but the feed is good even then and the next Wet turns it green again for the coming season.

About 21,000 head of cattle are run at Gogo and most of them are poll Shorthorns. About 4,500 are turned off each year to the meatworks at Broome to provide beef for the American market. The herd is free of brucellosis but there is a small incidence of tuberculosis which is being eradicated and which should have disappeared completely in a few years.

The property employs eighteen to twenty people in the working season and some of these come from the Aboriginal settlement which is on the property. Most of the time the property is worked with one stock camp of ten ringers and a head stockman, but other staff sometimes make up a second camp to work the nearer parts of the property. During the season they have to turn stock off at the rate of about 600 head every eight days and in order to do that the main stock camp works non-stop throughout the season.

They are helped in mustering by the station aircraft, a Piper Super Cub. It is a small two-seater and is fairly basic even for a light plane, but its manoeuvrability makes it ideal for the job. Max Ley takes off from the airstrip beside the station area to fly to the paddock being mustered. Still climbing, he passes over the Aboriginal settlement and steadies on a course for the paddock. Ten minutes later he can see the stockmen grouped together on their horses in a corner of the paddock as they wait for him to arrive.

He has already seen a small mob of cattle at a distant water and he decided to start with them. Communication with the men on the ground is cheerful and unsophisticated — he shouts through the open window of the cockpit as he flies low over them. They hear him first time and they pull their horses round to head in the same direction.

He gains height as he approaches the cattle so that he can assess the situation. He wants to move them towards the stockmen but watercourses or large hills sometimes make a direct route impossible. In that case he has to move them round these obstacles until he can take a straighter line across open country. Above all, he must never fly in front of them, for then the cattle will break in confusion and it will take a long time to reassemble them.

He banks the plane sharply now, then loses height rapidly to fly straight and low across the back of the mob. As they start to run he spins the plane round in a stomach-lurching turn and flies back as low as before. The cattle are moving as a mob now, but soon he will have to turn them to pick up a direct line to the horsemen.

He is barely 50 metres above them when he turns the plane and flies down the side of the mob, engine roaring as he passes. The leading beasts turn away and the others follow so that in a few seconds the mob has turned to run in the direction he wanted. Another screaming turn followed by a low pass directly behind them to make sure they remain on course, and then he climbs a little higher to watch them. They are moving well now and heading right for the stockmen who are riding towards them. He has given them twenty head in a few minutes.

The stockmen cannot see them yet but they have been watching the little red and white plane darting backwards and forwards, sometimes almost hidden by the trees, and they know the cattle are not far away. They spread out so that they can keep them together and Max Ley, satisfied that they have control of them, goes off for another mob.

This kind of flying needs a very good pilot. Although there are no overhead cables to restrict movement, there are plenty of traps for the inexperienced. If Max Ley were to turn the wrong way, for example, he could find himself looking straight at the sun and the temporary loss of vision when only metres from the ground can be the only mistake he needs to make. It is all a matter of time, speed and distance, and his judgement has to be right every time.

Aerial mustering is now so well established on many cattle properties that many younger ringers have never done it any other way. And those that have, like Len Hill, often wish that they had had the benefit of it in earlier times.

Len Hill still retains all the old skills of the traditional bushman. Born in 1926 at Mount Barker, he left school in Perth when he was sixteen and the following year went as a jackeroo to Bindi Bindi, near Walebing. From there he moved to Carnegie, near Wiluna in the north of Western Australia, then in 1946, when he was a stockman at Windidda, he helped round up a mob of horses and with a boss drover spent five weeks taking them up the Canning Stock Route to a station in the Kimberleys. They stayed to help with the muster and then took a mob of 550 bullocks back down the Stock Route. It took eighteen weeks to cover the 1,400 kilometres and it left him with a detailed knowledge of that route that few can equal now.

In 1949 he joined Vesteys as head stockman at Flora Valley, near Halls Creek, and after three years' absence in New Zealand he rejoined them as head stockman at Nicholson Station on the border of Western Australia and the Northern Territory. Shortly afterwards he became assistant boss drover and also acted as relief manager on other Vestey properties in the Kimberleys. In 1959 he was made manager of Nicholson and subsequently became supervisor of five other properties.

Rounding up a beast whilst camp drafting. Man and horse have to work as one for there is no time for the usual commands.

When the West Australian government resumed land from most of the Vestey properties in that State, the company decided to sell all its interests there and to concentrate on its properties in the Northern Territory. So Len Hill decided to leave the cattle industry and in 1980 he moved to Perth for a taste of city life. But the taste was not to his liking and when Tim Emanuel offered him the job of managing Christmas Creek he accepted without hesitation. In 1982 he took over as manager of the neighbouring property of Gogo.

One of his first innovations was the introduction of Africander cattle to the herd. When crossed with the poll Shorthorns he thinks they will produce cattle which will be ideally suited to this country. At the same time, he is trying to build a herd which will be as good as the legendary one that Vic Jones ran when he was manager of Gogo in the 1960s and which had a uniformity and general excellence that was famous throughout the Kimberleys.

'The only way is by ruthless culling, so that only the best are used for breeding, and by importing first-class bulls from other properties. It might take ten years, but by then we will have stud sales as well as good beasts for the meatworks.'

Like all good cattlemen he has an affinity with his beasts which is almost affection, but he never forgets the realities of running a cattle station. 'You have to temper humanity with good husbandry,' he says. 'Those that should be culled must be culled. The rest should be given the best you can provide.'

Before the flood he had set aside 260 stud cows at Jillyardie which were to be the foundation of his new herd. But by the time the flood water had receded there were only 90 left. He thinks they also lost about 3,000 head in the flood, but until all the paddocks have been mustered it can be little more than a guess. How do you cope with a situation like that?

'It's quite simple,' he says. 'You just start all over again. There are no short cuts in this business.'

The next morning the ringers are up early as usual. Their stock camp is near the yards and they have slept on the ground in swags with mosquito netting stretched above each one on short poles. They have breakfast under the awning at the side of the kitchen trailer, then mount up for a morning of camp drafting.

Although the yards are often used for drafting they have one disadvantage: cows and calves become separated and it takes a long time for them to find each other afterwards. Camp drafting, on the other hand, is done in the paddock and cows and calves are together all the time.

The cattle are taken from the yards and walked to a nearby creek. Unless they are fed and watered the younger cows will

Ashley Dickson, a contract bull catcher, in his stripped-down Landcruiser at Gogo.

The stock camp cook.

not mother their calves and the benefit of drafting in this way will be lost. When they have had a drink they are walked slowly across the paddock and held as a loose mob in one corner, with the stockmen forming an arc on the open side. Four men ride along the fence to the next corner and a fifth takes up his position half way along the fence. Then they are ready to begin.

The man doing the drafting walks his horse into the mob, in this case about 500 head, and walks slowly amongst the cattle until he sees a cow of the type being drafted. His job now is to work his horse so that this cow, together with her calf, is moved to the edge of the mob. Sometimes it is easy, but often it is not. As most of the beasts in the mob are moving as well it can be difficult to remain in contact with the selected cow. Some horses are particularly good at this and once they know the beast to be drafted they will stay in touch with it no matter where it might go within the mob.

When the cow and calf are on the edge of the mob the drafter keeps them moving until they are clear of it and standing alone. The ringers now move them down the fence to where the single rider is waiting to receive them. He then takes them to the other corner where the rest of the ringers are waiting for them, the start of a new mob.

Meanwhile, the drafter has walked his horse back into the mob to look for the next beast. This time it is not so easy for

when the cow and calf are at the edge of the mob they start to run. Other beasts join in so that suddenly this part of the mob seems to burst into splinters, with cattle running everywhere.

Some of the ringers spur their horses and gallop at full speed to head them off. The drafter does the same, for he is the only man who knows which beast he has selected. He tries to position his horse so that it is alongside the running beast and only a few centimetres from it. Then, still at full gallop, he turns his horse towards the beast so that it turns too. It doesn't always work. When the beast is fresh and full of running it might prop for an instant and veer off in another direction. Then he will have to catch it up and move his horse back into position so that he can try to turn it once more.

It is so fast that there is little time for the rider to command his horse. The horse must be able to work things out for itself and prop and turn in response to the movements of the beast. And the rider must be able to stay on whilst the horse is doing it — no easy feat when it is done at breakneck speed and the horse is wheeling round in turns that look impossibly tight.

Not all of them do stay on. One young ringer in a frantic chase across the paddock comes unstuck and crashes to the ground. He bounces slightly and then lies still.

A stock camp at Gogo. The men sleep in swags on the ground under a small mosquito net.

'Come on, get up,' yells the head stockman.

Another stockman catches the horse and leads it back, its reins mingled with his own. The man still does not move.

'Come on, get up. You think this is a bloody picnic or something?'

The man moves slowly until he is sitting up. Then he gets to his feet and takes the reins. It is often said that a rider who has been thrown from a horse should remount as soon as possible so as not to lose confidence. Some do, some don't. This man has no choice. Apart from the fact that there is work to do, it is the only way he would get back to camp.

He checks himself for damage and finds only bruises, although there are plenty of those. He stands by the horse's head whilst he sorts out the reins, then puts his foot in the stirrup and swings into the saddle. He winces as he settles down, then pulls the horse together and walks back to the others.

Len Hill rides to meet him.

'You're bound to come off if you ride like that,' he says. 'Your hands were going up and down and your arms and arse were going everywhere. Keep yourself tucked in and don't move any more than you have to. Go and take over from Jackie in the far corner and have a bit of a spell. You'll be right.'

As the man walks his horse away, Len Hill rides back towards the mob.

'Well, what do you expect — the Man from Snowy River? Let's get on with it.'

WALLAMUMBI

It is a sunny morning. The clouds of the previous day have cleared overnight and the early sun draws whisps of mist from the lake. In a paddock beyond the lake a group of horsemen are mustering cattle. It is quiet work, almost gentle, for these Herefords are calm and peaceful this early morning. The men ride easily, occasionally veering away to round up a beast which is not rebellious so much as lazy and needs to be reminded that it should stay with the rest of the mob.

It takes only a few minutes to muster the hundred bulls and soon the men are walking them from the paddock and along the road that runs beside the lake. A thin trail of dust hangs in the still air behind them, the sun flashes reflected light across the water and the big gums throw long and intricate shadows across the road.

The men lead the cattle towards a creek which flows into the lake and as the cattle slow down in hesitation a jackeroo shouts at them to send them across. He is told not to do it again. Shouting might add excitement to a nineteen-year-old's idea of dashing competence, but it does more harm than good at Wallamumbi. The mob splashes across the shallow water and the subdued jackeroo follows as the cattle wind slowly up the hill towards the yards.

There is hardly a sound now. The bulls plod quietly over the grass and even the dogs are content to trot quietly behind them. The man in the lead rides forward to open the gate and the mob passes slowly through, raising more dust now as they leave the grass for the bare earth of the yards. A jackeroo closes the gate behind them and the cattle mill around the yard as the men dismount and hitch their horses to the rails.

It is still early but there will be much to do today. It is a Friday in September, the day of the twenty-second annual Spring Bull Sale at Wallamumbi. The sun glints harshly on the corrugated walls of a large building on the other side of the yards. Empty now, it will soon be crowded with people. Then the bulls, together with others coming from a different paddock, will be sent one at a time into the straw-covered ring between the tiers of benches on either side and the building will echo the bids of the buyers.

But for the next few hours the men will be busy in the yards, drafting bulls into the order in which they will be sold, touching up the lot numbers painted on their rumps and making sure that none have developed any last minute defects.

David Wright, the owner of Wallamumbi.

Satisfied that everything is under control, David Wright leaves the yards and walks down the hill to the handsome new homestead that he built three years ago. He takes off his boots at the side door and pads through to the kitchen. His wife Margaret is already boiling the jug and soon they are standing side by side at the bench, enjoying a few minutes together before the hectic part of the day begins. Holding their mugs of steaming coffee, they look quietly at the country beyond the window, panoramic in its greenness, stretching away beyond the lake to the range of blue hills far away in the distance.

In front of the house a small paddock runs down to the edge of the lake and the sun, higher now, puts light round

the bullocks that are grazing there. Walking slowly, with their heads down, they nuzzle the grass with bovine thoroughness.

It is an idyllic picture and although David and Margaret Wright have seen it many times it still moves them. But for them it is more than just idyllic — indeed they barely recognise it as such — for it has a significance that goes far beyond the visual.

The rolling parkland was not always like this. Nor were the bullocks that add life to the foreground. At a glance they look like well-bred Herefords dwarfed by this extensive landscape. But when they are close to a fence which lends familiar scale they are no longer dwarfed. Suddenly they are huge, much bigger than Herefords usually are. It is their size that makes them significant for they are Beefmakers, and the story of Wallamumbi is about them and the fertile pastures that support them.

When David Wright's grandmother, Charlotte May, bought Wallamumbi in 1899 this country was far from idyllic. Fifty

years earlier the Maitland Chronicle had described Wallamumbi as consisting of 115,000 acres with a carrying capacity of no more than 2,500 head of cattle and in 1899 it was no better. Previous owners had tried to clear the dense forest that covered this country by ringbarking the trees but they had failed to control the regrowth and the country was, if anything, even worse than before.

Charlotte May Wright had moved from Queensland in the 1880s with her husband Albert and they had bought a property called Wongwibinda near Armidale. They still owned a number of properties in Queensland and they had moved south to escape the rigours of outback life and to establish a stud in this gentler country that would supply quality cattle to their northern herds. The stud was established with a few Herefords which were descended from five cows and one bull imported by the Wright family in 1827. These were the first Herefords to reach mainland Australia.

When Albert died shortly after the birth of their son Phillip, Charlotte May was urged to sell up. Although the improvements at Wongwibinda had been nearly completed,

The distinctive poplars of New England in the station area at Wallamumbi.

Phillip Wright, who started to run Wallamumbi for his mother when he was fourteen. Although he had no university education, he later became chancellor of the University of New England.

nobody at that time thought a woman could run a cattle property. Nobody, that is, except Charlotte May Wright.

Heavily in debt and with a young family, she rode out the depression of the early Nineties and continued to build up the herd of Herefords. As times improved she bought Wallamumbi and another property nearby called Jeogla and ran them both as an extension of Wongwibinda.

After a brief education at North Shore Grammar School, Phillip Wright returned at the age of fourteen to take up the running of Wallamumbi. Hard working and energetic, he slowly transformed it into one of the best cattle properties in New South Wales and the Hereford stud became famous throughout Australia.

But he did much more than that. Keenly interested in local affairs, he was instrumental in setting up the New England National Park and campaigned strenuously for a university to be established at Armidale to serve the needs of the northern part of the State. It was a successful campaign and when the University of New England received its charter in 1954 he became its first deputy chancellor. Later he became chancellor and in 1957 the University acknowledged his tremendous achievement by making him an honorary Doctor of Science.

But for all his varied activities, the development of Wallamumbi was his biggest concern. When he died in 1970 at the age of eighty-one the Bishop of Armidale said of him, 'He was a big man in stature, big in his vision and hopes and he reflected the challenge of a big country which he and his forefathers did much to tame, for he was a fourth generation Australian.'

His ashes were scattered in the paddocks of Wallamumbi where he had lived for sixty years — paddocks that now looked very different to those he had helped to clear as a boy.

His son, David Wright, had been born in 1933 and after being educated at The Armidale School he joined CSIRO at their research station in Armidale. In 1952 he went to Britain to gain experience as a jackeroo on Hereford properties there. 'At that time Britain was the only external source of Hereford stud stock. They bred bulls there for sale throughout the world and in those days we knew as much about English Herefords as we did about those in Australia.'

He returned in 1953 to do six months national service in the RAAF and then spent another six months travelling across northern Australia to examine cattle properties in the outback. He then returned to Wallamumbi and in 1958 took over the running of it from his father. His first concern was to continue the development of these pastures which had taken so much of his father's time.

The first attempts to improve the pasture at Wallamumbi were simple and crude: they burnt it. Then in the early 1920s, with Wallamumbi reduced to its present size of 12,000 acres, a more serious effort was made to increase the amount of useable land by ringbarking the trees.

A few years later exotic grasses were introduced on a small scale but they did not do well and had little or no impact, largely because the soil, like that of much of Australia, was deficient in phosphate and nitrogen. When this deficiency was recognised in the mid-1930s, superphosphate was used on a small scale with good results but the war intervened and postponed its use on a large scale.

By the late Forties there was much more enthusiasm for such developments. Whole paddocks were ploughed, treated with super and planted with exotic grasses, but because it could be done only at certain times of year there was a limit to the amount of country that could be treated this way. The answer was to do it from the air.

The first experiment in aerial spreading at Wallamumbi took place in 1951 with an old Tiger Moth. Bags of super were emptied into a hopper in the front cockpit, although the propeller blew much of it over the pilot sitting in the cockpit behind. The hopper had a valve in the bottom and this was operated by the pilot as he struggled to dodge the trees and fly low over the ground in his antiquated biplane. It was a hazardous business. But a year later they could see an improvement in the condition of the stock and four years later they could see the difference in the pastures themselves.

With the benefits of superphosphate now beyond doubt there was a rapid improvement in aerial equipment and this made it possible to work on a much bigger scale. At Wallamumbi the first big paddock to be treated was 1,500 acres and its carrying capacity increased dramatically.

By the mid-1950s the pasture was being improved so rapidly that it seemed impossible to stock it fast enough. As stock numbers grew, more money was needed to support them. Paddocks had to be subdivided into smaller ones, which meant more fences and more water; and more men were needed, which meant more houses, more horses and more equipment. But the benefits were real enough, especially when the exotic grasses responded to the increased nitrogen in the clover-dominated pastures. It all seemed little short of a miracle. Until the drought of 1965.

Suddenly the exotic grasses withered and died and the paddocks became empty and useless. 'The country looked dreadful,' says David. 'I thought I had ruined it.' The high stock numbers were now a liability and with no means of reducing them they had to resort to hand feeding to keep them alive. It was cripplingly expensive.

When the drought ended David Wright decided to plough the paddocks into seed beds and to sow a different range of drought-tolerant exotics that could not be sown from the air. But it was a mistake. The ploughing killed the drought-resistant native grasses and when drought returned in the late Seventies it was the unploughed paddocks with these better native grasses that supported the property.

'I realised then that the solution was to leave the paddocks alone and to introduce selected exotics to supplement the good native species. I shall probably never plough these paddocks again.'

The exotics are now introduced into the paddocks by sod-seeding. Grooves are cut in lines across the paddocks

John Weston. Once head stockman at Brunette Downs, he is now the manager of Wallamumbi.

The famous V2V brand, incorporated in this sign, was first registered by Albert Wright in 1872 and is now one of the oldest brands still in use by the same family.

Beefmaker bullocks in a paddock near the homestead.

Cattle buyers enjoy a barbecue lunch during the annual Spring Bull Sale at Wallamumbi.

and seeds of selected species are sown in them. Most of the turf remains undisturbed and the natives continue to flourish. The result is a blend of grasses which thrive in soil made more fertile by the use of super and which, because of their variety, can survive the harshness of drought.

In thirty years the paddocks of Wallamumbi have been transformed from infertile low carrying capacity country into the rolling greenness beyond the window. And if Wallamumbi now looks idyllic, it is because David Wright and his father made it so.

Whilst the improvement of pastures by the use of superphosphate was not unique to Wallamumbi — others in this area were doing the same thing at the same time — the development of the Beefmaker was.

As the pastures improved, so did the Herefords. But in time both reached a stage where it seemed little more progress could be expected. In some respects this was not unusual. Genetic gains are rarely constant and every breeder accepts that after a period of improvement his herd will 'stabilise' for a time and show little development until the next phase.

David Wright thought that this stability was more fundamental than that and that future changes were more likely to be a reflection of the seasons rather than real genetic gain. So, whilst keeping the Hereford stud intact, he looked for a means of improving his commercial herd so that they would become more efficient makers of beef.

A Hereford, like all breeds, reaches maturity at a certain age and after that any weight it puts on will be fat, not meat. As it takes about five times as much feed to produce fat as it does to produce the same weight of meat, the economic usefulness of the animal declines rapidly after it reaches maturity. So David looked for a means of deferring maturity so that the beast would continue to grow meat over a much longer period.

This could be done by the infusion of another, slower maturing, breed, but it was not quite as simple as that. The resulting mix would have to retain all the good characteristics of his present herd. It already had a remarkably high level of fertility. The cows were not only capable of calving in the paddocks without supervision, but they could also then provide abundant milk to bring the calf on. Furthermore, as the intention was not only to increase the production of beef but to sell breeding animals to other commercial producers so that they could do the same, the new cross would have to look like a Hereford to appeal to his existing clients.

With this 'blueprint' firmly in mind he decided, after much research, that the best breed for the purpose was the German Simmental. In 1973 he started to introduce imported semen into his commercial herd. After a period of trial and error, he settled on a cross of 75 per cent Hereford and 25 per cent Simmental, and the result was the Beefmaker.

Because it has the longer maturing characteristic of the Simmental, the Beefmaker continues to grow for about four years and during that time it will be significantly bigger and heavier than a pure Hereford of the same age. More importantly, the extra weight will be meat, not fat. Further, the new cross retains the appearance of the Hereford and has the same ability to calve easily without supervision.

Live weight, however, is a vital factor at Wallamumbi and is used as the means of selection in both stud and commercial herds. As the ability to put on useable weight is highly heritable, beasts selected for this ability will reliably pass it on to their calves and thus to the whole of the herd. Known as performance testing, it is the basis of the breeding programme at Wallamumbi.

It is in direct contrast to the more conventional system, known as progeny testing, in which outstanding bulls, either bought in or selected from the herd, are used repeatedly in the hope that their features will be passed on to the rest of the herd.

David Wright has little time for this method. He argues that not all the features of a bull are heritable, and those that are will be no more obvious in the twentieth calf he sires than they were in the first. The result, he says, is a diminishing gene pool which will have little potential for further progress.

In order to keep the gene pool as large as possible, he says, new bulls have to be used all the time. At Wallamumbi bulls are used for only one year. As this includes two mating periods each bull will sire two drops of calves before he is withdrawn from the herd and sold. There are few exceptions, no matter how outstanding an individual bull might be. Of the bull calves he has sired, some will be better than him and as they grow they will be used in his place.

Parallel with this, all the cows are expected to produce a calf each year and those that don't are culled from the herd. The result is a rapid progression of generations whereby the 'best sons of the best sons' are mated with fertile cows, so producing a genetic gain which is significantly faster than that obtained by progeny testing.

Not all breeders agree with David Wright and in an industry that is often slow to accept change he can understand why. But he is convinced of the benefits. 'If you select on the basis of growth rate and fertility and sell the bulls as soon as they have made their contribution instead of working them to death, then you must make progress. It is as simple as that.'

Until a few years ago Wallamumbi continued to be run in conjunction with Jeogla as one family property. Then in 1978 David and his brother Bruce decided to split the properties and go their separate ways. In the division, Bruce took Jeogla (which he still runs) and David took Wallamumbi, together with a 2,800-acre property called Woodburn which they had bought in 1965, and Achill, a 6,000-acre property which borders Wallamumbi which was bought in 1972.

The homestead at Wallamumbi, built in 1980.

Since then he has bought two more properties: the combined property of Forglen and Conningdale which consists of 6,000 acres not far from Wallamumbi; and the 6,700 acres of Yarrowyck about 25 kilometres west of Armidale.

Together with Wallamumbi they amount to 33,500 acres and although each of the properties has its own manager they are effectively run as one from Wallamumbi. There John Weston, once head stockman at Brunette Downs, is supervising manager of all the properties and has full control in David Wright's absence. For, like his father, David Wright's interests extend far beyond the boundary fence. He was Deputy Chairman of the Australian Meat Board for seven years and Chairman of East-West Airlines until recently; he is a member of the Executive of CSIRO, and currently President of the Australian Hereford Society.

At Wallamumbi they run about 12,000 head of cattle on the properties and in good seasons augment them with store bullocks bought in for fattening. There is a full time staff of twenty, including five or six jackeroos, and with families there is a total population of about fifty.

Wallamumbi itself is 48 kilometres east of Armidale and its 12,000 acres are divided into 84 paddocks. Most of it is pleasantly undulating country, but the area south of the Armidale-Grafton road is hillier and steeper. There is no underground water and the property depends on rainfall — an annual average of 33 inches. Three creeks cross Wallamumbi to provide permanent water in most seasons and it is one of these creeks which has been dammed to make the huge lake not far from the homestead.

Although it is a famous property its brand — V2V — is perhaps even more widely known. First registered by Albert Wright in Queensland on 8 June 1872, it is now one of the oldest registered brands still in use by the same family. Indeed, many cattlemen refer to their cattle not as 'Wallamumbis' but as 'V2Vs' and regard the brand as a hallmark of excellence.

That is one of the reasons the sale ring, so quiet and empty earlier in the day, is now bustling with nearly two hundred people. Some have travelled more than a thousand kilometres to attend this sale — some to buy, others to see for themselves the progress being made by the V2V brand. Now they are lounging on the benches chatting or studying the catalogue.

The gate at the rear of the ring is opened and a Beefmaker bull trots in. He stops in confusion and looks at the people in the raised stands on either side, then walks slowly to the far end of the ring. The buyers watch him as he moves, some

Moving a mob of bulls across a creek.

studying him with the assurance of experts who play with their own money, others acting for clients who depend on their judgement.

The auctioneer sits at a bench that runs like a bridge across the far wall at the end of the ring.

'Lot No. 75. A beautiful colour, ladies and gentlemen, and look at the strength in those hindquarters. A bull ready to go straight to work. Who will give me a thousand dollars?'

'Eight hundred.'

Men from Wallamumbi are standing round the ring facing the crowd, looking for bids. They are not long in coming and the men shout to the auctioneer as they see the nod of a head or the flick of a catalogue. Above them, the auctioneer starts his rapid chant which sounds like a rhythmic anthem.

'Nine hundred on my left — a thousand — against you sir — eleven hundred — twelve — thirteen — I'm bid thirteen hundred dollars, ladies and gentlemen — come on, this is a top class bull and quality like that is *never* dear — thanks Dick — I have fourteen hundred dollars on my right — fifteen — sixteen — seventeen — eighteen — I'm bid eighteen hundred dollars — nineteen — make it two thousand — two thousand I have — two thousand one hundred — two thousand two hundred — it's against you sir — it would be a pity to lose a bull like that for a hundred dollars — think of that terrible trip home without him — thank you — I have two thousand three hundred dollars...' He raises his arm above his head. '... for the first time at two thousand three hundred dollars — have you all done? — for the second time at two thousand three hundred dollars — have you all finished? — it's your last chance — I'm lowering the boom — at two thousand three hundred dollars for the third and last time ...' He slaps the bench. '... Two

thousand three hundred dollars to MacDonald of Charleville.'

The crowd relaxes and makes notes in the catalogue as the bull is taken from the ring and the next brought in.

'Lot No. 76, ladies and gentlemen. Who will start me with fifteen hundred dollars ...?'

By 3.30 in the afternoon the auctioneer is offering the last bull: Lot No. 113, a pedigree Hereford bull called Wallamumbi Byron. It fetches $1,200 and the sale is over. The buyers and their families clamber down from the benches and stream out into the bright sunlight. Some walk across to their cars, others walk round to a nearby building that has been set up as a bar and tea room.

Country men are soon in earnest conversation with each other, talking about the sale, the weather, or their own properties, whilst their wives talk of other things. Children race round outside until they are bundled into cars that speed off down the drive to the bitumen that leads back to Armidale. And as the light fades the men from Wallamumbi come in from the yards where they have spent the day marshalling the cattle. Now they get a beer, roll cigarettes, and talk about the day.

A jackeroo walks across to talk to Bill Adam, the overseer and stud manager, and points to a truck loaded with two bulls which is pulling slowly on to the road.

'Bet you feel a bit proud Bill, don't you? Seeing bulls like that leaving the place?'

'Yes I do. And I'll tell you what makes me even prouder — that's seeing a jackeroo leave this place who knows what he is doing.'

'I can only try.'

'You'll be right.'

THE COMPANY BOSS

Arthur Bassingthwaighte was born at Toowoomba, Queensland, in 1919 and was brought up on the family farm on the Darling Downs. There, he learnt to ride when he was three, was droving with his father at four, and at six was riding 8 kilometres a day to the local State primary school.

He was thirteen when his father died and being one of five children he left school to run the family dairy for six months before being sent to Southport School for two years. He left when the family property was sold and at fifteen went to work for two of his elder brothers who had a cattle property near Chinchilla. With the brothers away for much of the time, he worked with a deaf and dumb station hand and together they put up 5 kilometres of split post fencing.

He left in the drought of 1936 to work on a property near Taroom, then when the drought broke he took on a droving job. Helped by only one man, they drove a mob of 250 spayed cows over a distance of 390 kilometres. It took three weeks and at one stage they had to cross 67 kilometres of dry country that had neither water nor yards.

After that he worked on Auburn Station and at seventeen he was earning £2 a week plus four pence a head for spaying cows. He then did another droving trip, this time with a mob of a thousand cows and calves, before joining a property near Injune that ran 10,000 sheep and 2,000 head of cattle. Eight months later, when he was eighteen, the owners offered him the job of head stockman on another of their properties. Called Crystal Brook, it consisted of 450,000 acres on the headwaters of the Maronoa River in the rugged foothills of the Carnarvon Ranges. It ran 12,000 head of cattle and the nearest fence beyond the horse paddock was the boundary fence 65 kilometres away.

In 1941 he joined the AIF as a private and saw active service in a Commando Unit before being demobbed in 1946 as a Captain. Having married during the war he now bought a small farm outside Brisbane and started a trucking business as well, but soon he could see little future in either and sold them both in 1950. He joined the Primary Producers Co-operative Association and,

Arthur Bassingthwaighte, Chairman and Managing Director of King Ranch, Australia.

having studied accountancy in his two years at boarding school, went to the Charleville branch as bookkeeper.

In 1954 he joined Swift Australian Company, a subsidiary of the big American organisation, in Townsville, as assistant to the manager and in charge of buying livestock for their meatworks. Three years later he moved to Gladstone as livestock buyer for the whole of Queensland and then moved to Brisbane to become responsible for procuring livestock from all over the country.

He was made a director in 1960 and two years later, he organised with King Ranch a joint take-over of the Queensland National Pastoral Company, which owned seven properties in Queensland and the Northern Territory. He became managing director of the new operating company and the following year he was made a director of King Ranch and asked to take over the management of Swift Australia. He was the first Australian to run that company and after much reorganisation he was able to turn it into profit.

In 1967, when Swift's had problems overseas, they sold the Queensland part of the operation to F. J. Walker. A condition of the sale was that Arthur Bassingthwaighte continued to run it, which he did. In 1976 he was made an Officer of the Order of Australia for his services to primary industry.

In 1977, having been an active director of King Ranch for fourteen years, he was asked to take over its Australian management. At that time King Ranch had ten properties spread over New South Wales, Queens-land, the Northern Territory and Western Australia and they were incurring heavy losses. He embarked on a process of reorganisation and by 1980, as Chairman and Managing Director of the company, he had reduced the number of properties to nine, all in Queensland, and which together ran about the same number of cattle as had the original ten.

Tall, lean and with thinning grey hair, he believes strongly in personal integrity. 'You should never let anybody have anything on you. If you do, it is bound to catch up with you sooner or later and you will be forced into a position of weakness and compromise.'

He now lives and works in Brisbane, although he visits each King Ranch property four or five times each year. And his hobby? Running his own 11,000-acre 'weekender' near Brisbane.

INNESVALE

'OK, you blokes. Daylight!'

It is nothing of the sort. It is 5.30 in the morning in the stock camp at Innesvale and it is still dark enough to count the stars. But if the boss says it is daylight, then it is.

As men get up from their swags in the darkness they are like forms rising from the ground itself. Then torches come on as they make their way to the river for a cold wash. The cook throws shadows across the clearing as he moves round his early morning fire. Cigarettes glow briefly, men cough as they struggle into their clothes, and from beyond the trees comes the jingling sound of the horse tailer rounding up his mob.

By the time the cook bangs his tin tray to announce breakfast the sky is becoming lighter. The men walk over to the trestle table and watch him lift the cooking pots off the fire. Then they help themselves to either last night's beef or this morning's pancakes, cooked in batches in the camp oven and served with sweet syrup, thick and brown. There is a drum of tea on the iron frame which stands over the wood fire and the men fill their tin mugs from it before sitting wherever they can, talking quietly as the first light trickles through the outstretched fingers of the pandanus palms.

One man, slightly older than the rest, finishes first and makes his way to an open vehicle that looks like a jeep but which is actually a stripped-down Landcruiser. In the back is a communication radio and as he switches it on he glances at the nearest tree to make sure that the wire aerial he threw into it the previous evening is still there. He turns the dial until a small torch bulb glows at its brightest, then picks up the microphone and presses the button.

'Mobile to Innesvale.'

'Yes, mobile, this is Innesvale,' says his wife.

'Everything all right Kay?'

'Yes, all OK. Jim didn't ring back last night but the light on the radio phone was on all the time so he might not have been able to get through.'

'Yeah, he'll probably ring this morning. Kay, it looks as if we might have these cattle into the yards sometime tomorrow if we don't hit any snags. Would you ring the meatworks and see if they can take them about the end of the week?'

'Yes, all right then. Do you want me to bring out any more Avgas?'

'No thanks, we should have enough if we don't get held up.'

Ian McBean, the owner of Innesvale. Once a drover, he is now the owner of the biggest private block of land in the Northern Territory.

'OK. Talk to you this evening then.'

'Yes. Look after yourself. Mobile out.'

They have not seen each other for over a week, but as their conversation might have been heard anywhere between Alice Springs and Darwin it is not the best time for tenderness. Thirty kilometres away in the small office at the Innesvale homestead Kay McBean puts down the microphone and thinks of all the things she has to do that day, whilst out on the stock camp, now touched by the sun, Ian McBean starts packing up the set.

At fifty-two, Ian McBean is short and nuggety and looks durable enough to last for ever. He has that distinctive

177

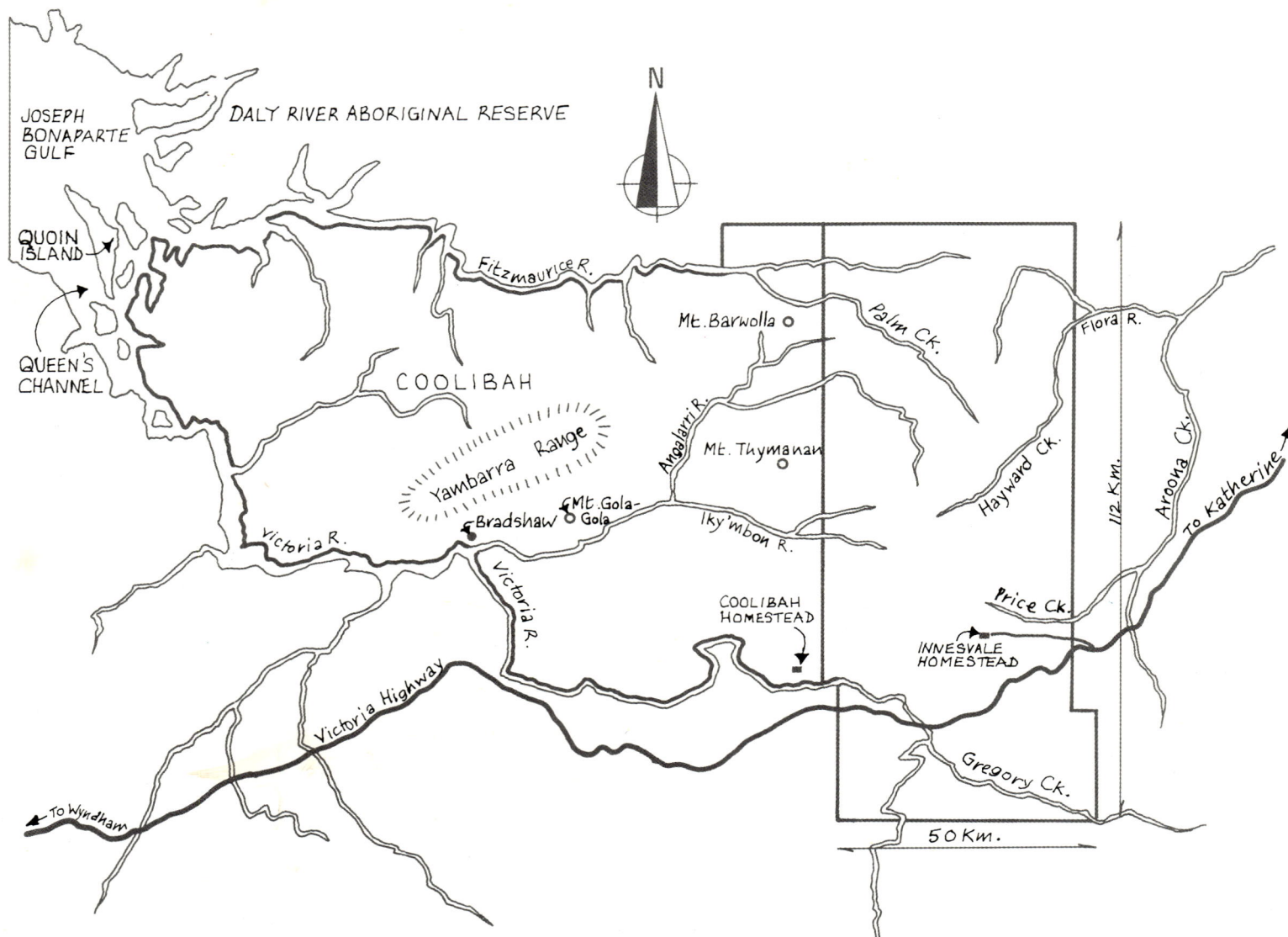

toughness that some men develop after a lifetime in the Australian outback and which others often mistake for harshness. It is a toughness born of the need to survive both physically and financially in an environment which is not conducive to either. But it produces more than toughness: it produces determination, understanding, and a compassion for those who are not surviving so well. In some it produces harshness too, but not in Ian McBean.

Because of this toughness, this air of knowing exactly what he is about, most people would recognise him as the boss as he walks back to talk to the men in the camp. In this part of the Northern Territory he could be an experienced manager running the property for some distant owner. But Ian McBean is not the manager — he is the owner of Innesvale. Indeed, he is the owner of the biggest private block of land in the entire Northern Territory, although the description embarrasses him. 'I'm just a busted-arsed drover,' he says. 'Like I always was.'

In part, at least, it is true. Twenty years ago Ian McBean was a drover working out of the small town of Camooweal on the border of Queensland and the Northern Territory. His job was to move mobs of cattle overland from one station to another, or from a station to the meatworks. As a self-employed contractor he took jobs wherever he found them for the best price he could get. In good seasons, when jobs were plentiful, he made a reasonable living. But at other times, perhaps most of the time, it was a hard life that brought little reward.

There were two kinds of drovers. One kind was hungry for money and tried to do each job quickly so that they could get on to the next. They pushed cattle along at a cruel rate and by the time they delivered them they were often in much poorer condition than when they had accepted them. 'Drivers', they were called by people like Ian McBean and others who did the job differently. For these people took pride in looking after the cattle and instead of pushing them along day after day, kilometre after kilometre, they let them walk along at their own pace. They might do only 12 kilometres each day, but with good feed and attention to water the cattle would hold their condition and might even improve on the way.

Although well educated, Ian McBean learnt his bushcraft by experience, for there is no other way. Once, when he was a young man working for an experienced boss drover in the

Moving the coachers during a two-week muster at Innesvale.

Dinner camp.

Queensland Channel country, one of his duties was to light a smoke fire so that the horses could stand near it to escape from the millions of flies. One day he found a big hollow tree and thought it would be an ideal place for the fire.

He had a good one going by the time the boss drover rode into the camp and expected a word of praise for his cleverness. Instead the boss told him roundly that it was a stupid thing to have done because some time during the night the tree would burn through and fall. That would certainly spook the cattle, which were held in an open camp, and they would rush in all directions. He was told to stay on watch until the tree fell and to be ready for action when it did.

It was not until 3.30 in the morning that the tree finally crashed to the ground in a great shower of sparks, and the cattle did not even get up, let alone bust. But he had learnt something about lighting fires, and once was enough.

Later, when he was a boss drover himself, he was booked to overland a mob from Coolibah Station. But when he arrived the mob was not ready, so instead the owner offered him some contract-mustering on an outlying part of the property called Innesvale. He did it, was impressed with the country, and thought no more about it.

Then in 1964 he was taking a mob east from Auvergne Station and was camped at the small settlement of Timber Creek on the Victoria River when he was joined by a mate travelling in the opposite direction. His mate told him that Innesvale had been resumed from Coolibah and was available for selection by ballot.

It was a familiar situation at that time. The big pastoral properties were held on leases from the government and if they thought the property was not being fully used they would reclaim part of it (or in extreme cases all of it) with a view to letting another grazier take over. The terms were reasonable and the number of applicants was usually considerable. Those who had sufficient experience and capital to run the place went into the ballot, and the new owner was decided in a draw.

Talking well into the night, the friend tried to persuade Ian to apply. He might have no money, to be sure, but he had assets in the horses and equipment that made up his droving plant and that would give him a start. It all seemed slightly unreal to a 'busted-arsed' drover, but Ian finally decided it was worth a go. He telegraphed the owner of the mob for permission to leave it in the care of his off-sider, and when this was given he headed off to Darwin to file an application for Innesvale.

There were another fourteen ballots open at that time and most applicants entered them all. But Ian knew that Innesvale carried enough unbranded cattle to give him a good start, and that was the only one he applied for.

Weeks later, when he had almost forgotten about it, he was in the pub at Camooweal when a man rushed across the road from the small post office waving a telegram and shouting the news. Ian McBean was the new owner of Innesvale.

The homestead at Innesvale.

In the station yards.

Innesvale at that time was not a property at all, it was simply 4,300 square kilometres of virgin tropical bush. There were no improvements of any kind: no buildings, no fences, no yards — nothing. He could not even get a vehicle in and when he arrived for the first time it was with one packhorse and lots of determination.

Step by step he started to improve the place. He built yards so that he could brand and turn off the cattle he could muster. He built fences so that at least some of them could be kept out of the wild, untouched country. And in time he built a big single-storey homestead for Kay and the kids. He was always broke, for whatever money he made was immediately allocated to more improvements.

He was on a financial knife-edge for much of the time. Cattle prices were low and the remoteness of Innesvale meant that high freight charges inflated the cost of everything. It was pioneering in the true sense — with few comforts, no security, and no guarantee of success. As the property changed, so he changed with it. He learnt how to run a business, where to find money, and how to deal with people in the city. And he survived and Innesvale became a station.

Indeed, by 1982 it was a fairly prosperous one, although the living was still far removed from the popular concept of a grazier's life. He could not afford to employ many people and

Moving the stock camp. The trailer, called the tucker cart, carries the food and cooking gear. On top are the rolled swags of the stockmen.

he was still working as his own head stockman, but by now he knew he could see himself through a run of bad seasons. Then, out of the blue, he heard that Coolibah, which adjoined his western boundary, was for sale.

By the time he heard of it, it seemed that it had already been sold to overseas buyers, although the deal had not yet been completed. Ian McBean now flung himself into action of a different kind. Assuming that he could raise the money somehow, he started on what was an almost continuous journey from Innesvale to Canberra to Darwin and back to Innesvale, then round again. He pestered politicians and public servants, he talked to banks and lending institutions and occasionally, when he had time, he talked to Kay. Then after months of negotiations he finally bought Coolibah, and he was broke once more.

The two properties cover some 14,245 square kilometres of natural tropical country 176 kilometres west of Katherine. The Fitzmaurice River makes up most of the northern boundary, whilst the Victoria River forms the boundary to the south. The property extends west to where these two rivers flow into Joseph Bonaparte Gulf, some 170 kilometres from the eastern boundary of Innesvale. It is hilly, at times mountainous, country and is crisscrossed with rivers and innumerable watercourses that run high in the Wet. It has a rainfall of 38 inches a year and the growth is natural, lush

and prolific. The rainfall is sufficient to provide adequate water for most of the country, but there are seven bores at Innesvale and tanks at Coolibah catch the spring water that runs off the escarpments.

There are only twenty paddocks in the whole of the property and most of these are close to the station areas of Innesvale, Coolibah and its outstation at Bradshaws. Cattle can be run in fairly controlled conditions in the paddocks, but the rest of the property is a vast area of open range country and the cattle run wild.

Most are indigenous Shorthorns together with Brahman bulls and they have produced practically every possible combination of Shorthorn-Brahman cross. There are about 20,000 head, although with the nature of this country it is really anybody's guess. About 2,000 head are turned off from Innesvale each year and they hope to double this when Coolibah is in full production. He cannot afford to restock Coolibah but there are some good Brahmans in the homestead paddocks, which will form the nucleus of his breeding programme.

Ian now employs about twenty-five people during the working season, many of whom are Aborigines. With families, the total population is about fifty. But he still prefers to be his own head stockman and to work the cattle himself. This is partly because he doesn't think he can afford one, and partly because with all his experience he thinks he is the best he would find anyway, especially for cattle as wild as these.

In his opinion there is no substitute for experience when handling cattle and he regrets that many young stockmen are now unable to get much. This counts in small things that others might not notice. For example, if he is moving a mob along a track that passes a gully, he will place a man there so that the cattle don't try to rush into it. A less experienced man might not anticipate that possibility and if the cattle did break, the ringers would have to spend valuable time rounding them up again.

He shares his knowledge and experience freely with his men, provided they are worth the effort! The working season is spent in a non-stop cycle of mustering, yarding, branding and turning-off until the Wet starts and the work has to come to an end. When it does, most ringers know a lot more about handling cattle than they did when they arrived.

At the start of the season they muster the homestead paddocks on horseback so that they can turn some cattle off fairly quickly. These cattle have been worked before and so are relatively easy to handle. But once the paddocks have been mustered they have to go out into the far country where the ground is rougher and the cattle are wild, and then they must use a different technique.

After leaving the station area, they muster the first enclosed paddock. They have mustered it before so they do not expect to find many. Indeed, at the start of this muster they found only forty beasts, but it was enough. This small mob, known as the coachers, is then walked to a distant part of the property where the real muster is going to start, which in this case was some 100 kilometres from the homestead. During the walk these cattle become used to men on horses and by the time the muster starts they are quite 'educated'.

Having reached the far country the men turn the coachers round and start to make their way slowly back, except that now they muster the country as they go. The cattle they find join the coachers and the mob gets bigger each day. The coachers could almost be called coaxers (and perhaps once were), because they coax in the wild cattle flushed from the hills and gullies. These cattle might never have seen a man on a horse before, but they are willing to join a mob of peaceful cattle who seem unconcerned about it all. This technique is as old as cattle in Australia, except now it has a modern refinement: the wild beasts are flushed out by a helicopter instead of men on horses.

Now, as Ian McBean walks back through the stock-camp, they have been mustering with the coachers for two weeks, day after day, slowly working their way back to the yards at the station. In the process the mob has grown from the original forty to several hundred head and with luck they will have them in the yards tomorrow.

It is light now and men are rolling their swags and dropping them beside the tucker cart so that they can be taken to the next camp. The horse tailer, who got up with the cook at 4 a.m., has rounded up the horses and he and his

Stockmen move the mob across the river at the end of a day's muster.

mate have given each one a nosebag feed. Some of these horses will be used today by the ringers whilst the rest, including the packhorse that carries the midday meal, will be walked along behind the cattle.

Ian rolls his own swag, then goes through the trees into the next clearing where the frail blue helicopter has been parked for the night. The pilot, dressed in thongs, shorts and a bush shirt, is polishing the glass bubble with a cloth, his helmet sitting in readiness on the seat inside. He works for a charter firm in Darwin and flew the machine down for the start of this muster. Since then he has lived with his swag in the stock-camp and each day blended his modern skills with the more traditional skills of the horsemen on the ground. It is a blend that nobody finds curious anymore.

They talk together beside the machine and Ian describes the route he will follow that day and the places along it where he can hold the coachers at rest to receive the beasts the pilot will drive out of the scrub. They talk of the country the pilot will cover and Ian draws a map on the ground to show him where the creeks and gullies are, for that is where he will find the cattle.

By the time Ian returns to the camp some of the men are mounted up and moving out. He collects his horse from the tailer, as the cook gets ready to pack the tucker cart so that it too can move on.

The cattle have been held for the night in a temporary yard built of tubular steel. Some of the men are standing by to dismantle it so that it can be taken to the next camp, where it will be erected again to receive the mob at the end of the day. Meanwhile, the mounted men take the cattle to the river for a drink before tailing them for a while on good feed. Then they will move them off for the start of the day's journey.

Back at the camp the pilot buckles on his helmet and climbs into his seat. He carries out his pre-flight checks then he starts the motor and the blades overhead begin to turn. The motor runs more evenly, the blades turn faster and he lifts the machine smoothly from the ground and curves away above the trees. He gains height, then kicks the tail round to head for the nearest creek about 16 kilometres away.

It is pleasant flying. The sun is warm and the air is smooth and clear. The blades are an invisible blur above and the motor behind him has set up a rhythm of its own as the ground slides past beneath him.

In a few minutes he sees the winding course of the creek and he loses height before turning to fly along it. The ground is well timbered and soon he is only a few metres from the canopy of the trees, looking for cattle in the clearings or along the water's edge. They are not easy to see, especially when they are lying down in the shade of the trees, which is what they are likely to be doing now.

He sees a small group on the edge of a clearing. He swings the helicopter round and skids back, almost touching the trees. Then he lowers the machine into the clearing until he is just above the ground and flies across the cattle. They get to their feet and stand for a moment in confusion before

This bull, which has been caught by Ronnie Ogilvie, is too wild to join the mob, so it is winched into the truck to be taken straight to the meatworks.

trotting off in the direction of the coachers. Although they cannot see them, or even know they are there, his manoeuvre has skilfully headed them in the right direction. He spins the machine again and lifts it clear of the trees. He makes a couple of passes over the mob to make sure they keep going, then climbs away from them to look for more.

He finds another group not far away and soon has them running to join the first mob. But a wild, unbranded bull at the back of the group is not as frightened as the rest and, propping and turning, it veers off into the scrub. He banks the machine in a tight turn and goes after it. The bull is galloping hard through the trees, often hidden by the canopy, but the pilot anticipates the direction it is taking and manages to stay in contact with it, although he cannot yet turn it.

It is fast and dizzying. The bull keeps turning past the trees, crashing through the undergrowth and jumping across the small gullies that run into the creek. Above, the machine reflects the same movements, spinning, turning, often tail up so the pilot can look directly down at the ground, or banking so that he can look through the open sides of the bubble.

The bull rushes into a clearing and looks around. The machine comes down behind him until it is only a few metres from the ground and the pilot edges it forward until the skids are just behind the rump. Then the bull turns, puts his head down and charges. The pilot whips the machine away and turns a full circle with the blades barely clear of the ground. When he comes round the bull has gone. He curses and climbs out of the clearing, hovering so that he can examine the surrounding trees. He cannot see the bull and assumes it is hiding in a small gully. If so, it is not likely to come out again whilst the helicopter is overhead.

He breaks off and flies back to his mob. The cattle have slowed to a trot now but they are still going in the right direction. He flies over them to keep them moving, zigzagging backwards and forwards behind them, sometimes flying the machine sideways to avoid passing in front of them. Then he climbs higher to locate the coachers. They are

Heading towards the station yards at the end of a two-week muster at Innesvale.

almost directly ahead and about 3 kilometres away. A huge mob of red and white cattle, they are at rest in a large clearing with the stockmen on their horses positioned at intervals around them.

He flies back to his cattle and pushes them along once more. That should do. He climbs again and sees a group of stockmen ride towards the edge of the clearing. They have been watching the helicopter and know that he is bringing in some cattle. The pilot, almost stationary now, watches the cattle run through the trees and into the clearing. They stop as they see the stockmen, then trot forward to join the coachers and blend in instantly, secure in the camouflage of numbers.

All but one. A huge unbranded Brahman bull with horns more than a metre wide bounds from the side of the mob and gallops at full speed towards the trees. Three stockmen pull their horses round and charge after it, trying to keep it in sight as it races along. Further back, Ronnie Ogilvie guns the motor of the jeep and tears after them, bouncing across the ground as he drives into the trees after the bull.

The ground becomes rougher now. Long grass conceals boulders and fallen logs and the jeep leaves the ground as it careers over them, bucking and rearing at suicidal speed. He can see the bull ahead. It is still going strongly, darting through the trees and swinging round them with the lithe, rippling motion of a superbly fit wild animal.

Ronnie Ogilvie follows it, steering a course through the obstacles, pushing down small trees with the vehicle, spinning round the more substantial ones. Branches crash over the vehicle and the grass showers it with clouds of seeds which at this speed hit like shotgun pellets.

For a while the bull has most of the advantages: it can turn more quickly than the vehicle and it can jump deep gullies which the jeep has to cross more slowly. But the vehicle has one overriding advantage which will finally tell: it can keep going longer.

At first it is hardly noticeable, but then it is clear that the bull is not running quite so fast. Ronnie Ogilvie brings the vehicle bouncing and jumping alongside the still galloping

animal. They are only centimetres apart now as they crash through the trees side by side. The rump of the bull is close to his elbow and one of the horns is over the side of the bonnet. Then he edges the vehicle against the shoulder of the bull and nudges it sideways. The bull breaks its stride, recovers, and he does it again.

The bull slows down, veers away, and then turns to charge the front of the oncoming vehicle. Ronnie hits the brakes and tries to stop the vehicle before the bull hits it, but even then they meet head on at what seems an incredible speed. The bull staggers and turns and Ogilvie spins his vehicle to ram the beast along its flank. The bull falls to the ground and is held panting beneath the tyre-festooned bullbar.

Ogilvie jumps out and ties the back legs together with a leather strap as the bull snarls at him. He is reversing the vehicle away from the bull as the three stockmen ride up, their horses blowing after their own spirited chase. They had kept going in case Ogilvie was unable to keep up with the bull in the rough scrub, in which case they would have taken over from him. At the end of the chase one of them would have tried to throw the bull over by twisting its tail.

Now, one of the men dismounts and takes a saw from the vehicle to cut off the needle-sharp tips of the bull's horns, tips which would do untold damage to other animals in the close confinement of a road train or the yards. They decide that this bull is too wild to go back into the mob so they put a rope round its horns and tie it to a sturdy tree. Later, they will bring up a truck and winch the bull into it, where it will buck and crash all the way to the meatworks.

It goes on all day and it is 6 o'clock before they have the mob in the newly erected yards near the night's camp. The helicopter pilot has pushed about fifty beasts into the mob during the day and Ronnie Ogilvie has six bulls secure and ready to be loaded. The stockmen have nursed the coachers along, gathering in the new arrivals and occasionally dashing off after a beast who has decided that its future might be more assured elsewhere.

The light is fading as they unroll their swags on the ground in the new camp and make their way down to the river for a clean up. By the time the cook bangs his tin tray the setting sun is turning the whispy clouds a vivid red and the sky is turquoise between them.

After the meal, men lie on their swags talking quietly or reading by torchlight whilst Ian McBean makes his call to Kay on the radio. They will be home tomorrow. When he has finished he has a yarn with some of the men by the fire, now sending flickering shadows across the grass. Somebody puts a Slim Dusty tape on a cassette player and his unmistakable Australian voice blends with those in the camp.

The tape finishes, torches go out, cigarettes disappear and at 8.30 the camp settles down for the night. The men get under their blankets and look at the stars above before giving themselves up to the dark and silence of the night. Not quite silent though, for from the nearby yards comes the soft sound of cattle, joining in a sort of communal snoring that will ebb and flow throughout the night.

A combination of old and new skills. Ian McBean has a conference with the helicopter pilot who will flush wild cattle out of the tropical bush.

'OK, you blokes. Daylight. Let's go home.'

It will be an easier day. They will be crossing country that has been mustered before, so there will be no more cattle being flushed out of the scrub to join the mob. The pilot will fly the helicopter to Darwin so that it can be serviced before he flies back to join the next muster in a few days time. The yards and tucker cart will go back to the station area as soon as they have been packed up, and the stockmen will gently push the cattle along the 10 kilometres to the yards. It will take most of the day, but it should not be too demanding.

By late afternoon the cook is preparing the evening meal in a real kitchen for a change and men are unloading the vehicles in the maintenance area. Beyond the station area the huge timber yards stand empty, waiting.

Then there is the sound of cattle in the distance and the muted shouts of the men as they move them along. Below the yards a horseman rides out of the trees, walking his horse and sitting easily, turning his head from time to time to make sure the cattle are following. Behind him comes the head of the mob, a file of cattle plodding along just as easily.

When the lead starts to climb the rise to the yards the cattle behind are still coming out of the trees. There are more men too, walking their horses beside the cattle and moving at the same easy pace. Then the tail of the mob comes through the trees and behind them a handful of stockmen spread out as they reach the open country. And the whole column of cattle, now a kilometre long, winds its way up the hill towards the yards.

The noise is greater now. The cattle are lowing, some bellowing, and the whistles and shouts of the men are more distinct. The procession seems almost triumphant, for it is the end of a fourteen-day muster through some of the most rugged country in Australia. And this group of men who started the first day with forty head of cattle are now bringing in a mob of eight hundred. If this were a Hollywood movie, and in truth it looks just like one, there would be crashing fanfares and stirring music as the men bring these cattle up the hill and turn the lead towards the gate.

Instead, there is dust. It rises in clouds as the men in front take up their position on each side of the gate and cattle jumble their way through into the yards. The men who were riding alongside the mob form a wide corridor that leads to the gate and the cattle walk endlessly past, snorting and swishing their tails as the dust swirls around them. Then, as the men stand in an arc round the gate, the last of the mob goes through.

A man dismounts and closes the gate. Then they all dismount and stand by the rails, looking at the cattle, appraising them, recognising the ones which gave them trouble. Ian McBean looks at them too. He knows that in order to keep this huge property going he has to bring in a mob like this every few weeks, working non-stop throughout the season which will have no weekends. It is hard physical work, work which he has done many times before and which he will do many times again. Routine.

But I can still hear the fanfares.

TULLY RIVER

It is 6.30 in the morning. The first light is glowing in the eastern sky and in the distance the mountains separate themselves from the night and take substance from the lightening sky. The sun appears and hangs briefly, vividly, on the horizon before it climbs higher and slides behind a bank of cloud.

Light now picks out the skyline of the western mountains and darkness recedes down the slope towards the valley. The valley itself is covered in mist, ethereal, tinged by the colours of the dawn. Then the mist, too, starts to retreat. Slowly at first, revealing the tops of the trees so that they stand out darkly against the streaky greyness that surrounds them. In the higher foreground the palm trees are more defined, their fronds hanging motionless in the still air.

For a few moments it is like a Chinese painting. The foot of the mountains is streaked with mist whilst lower down the trees are only suggestions in a monochrome landscape.

But now the sky starts to glow with streaks of red and purple as the sun slides out from behind the clouds. The mist dwindles to a few resilient streaks, the mountains solidify against the sky and in the valley the greyness turns to a vivid green that is startling in its suddenness. It is no longer a Chinese painting. It is the start of another day in tropical Queensland.

Tully River Station is 33 kilometres from the coastal town of Tully and 140 kilometres south of Cairns. As the light grows it picks out the high ground of Hinchinbrook Island not far offshore, whilst a little further north the early rising

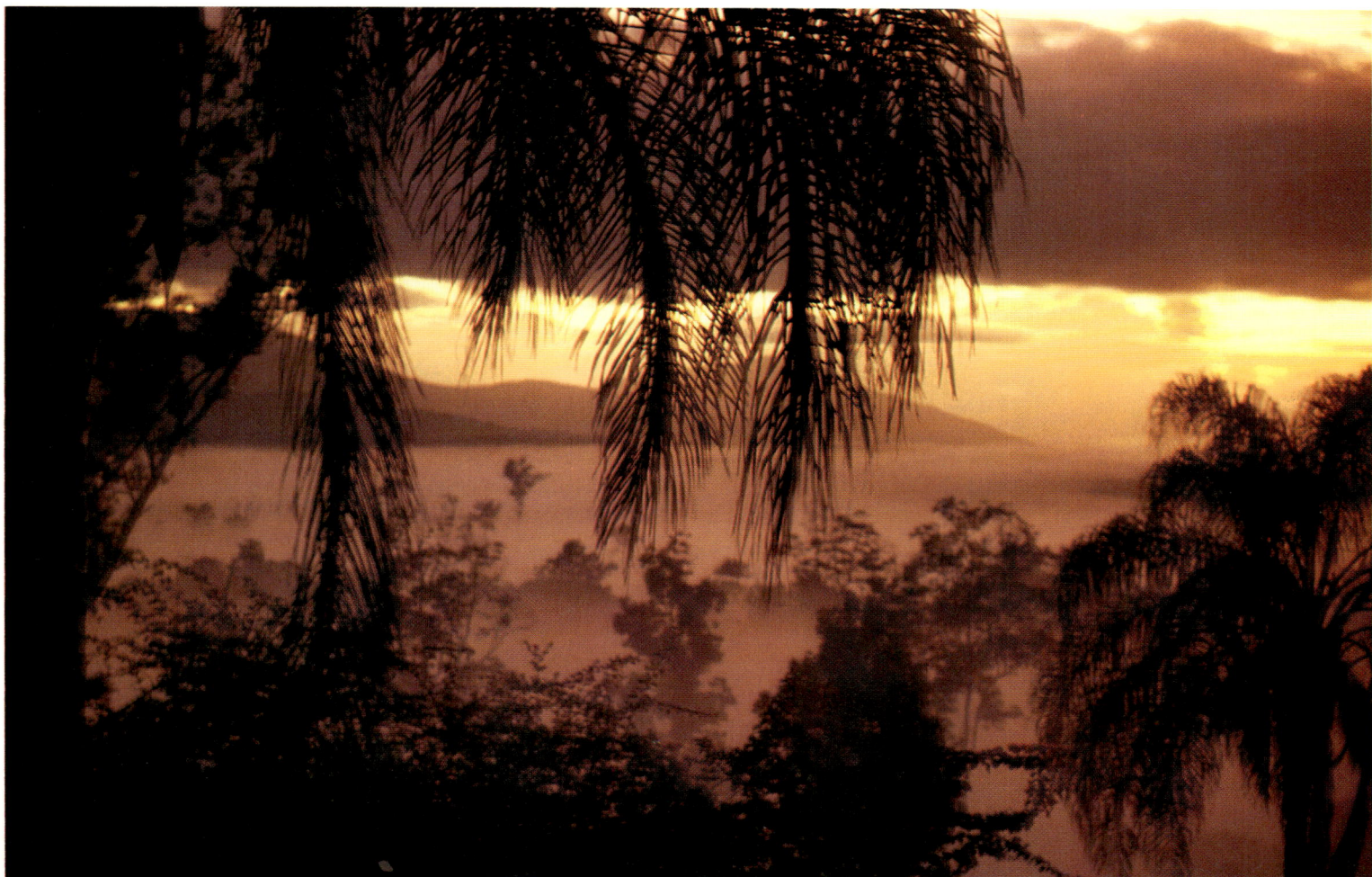

Dawn mist at Tully River.

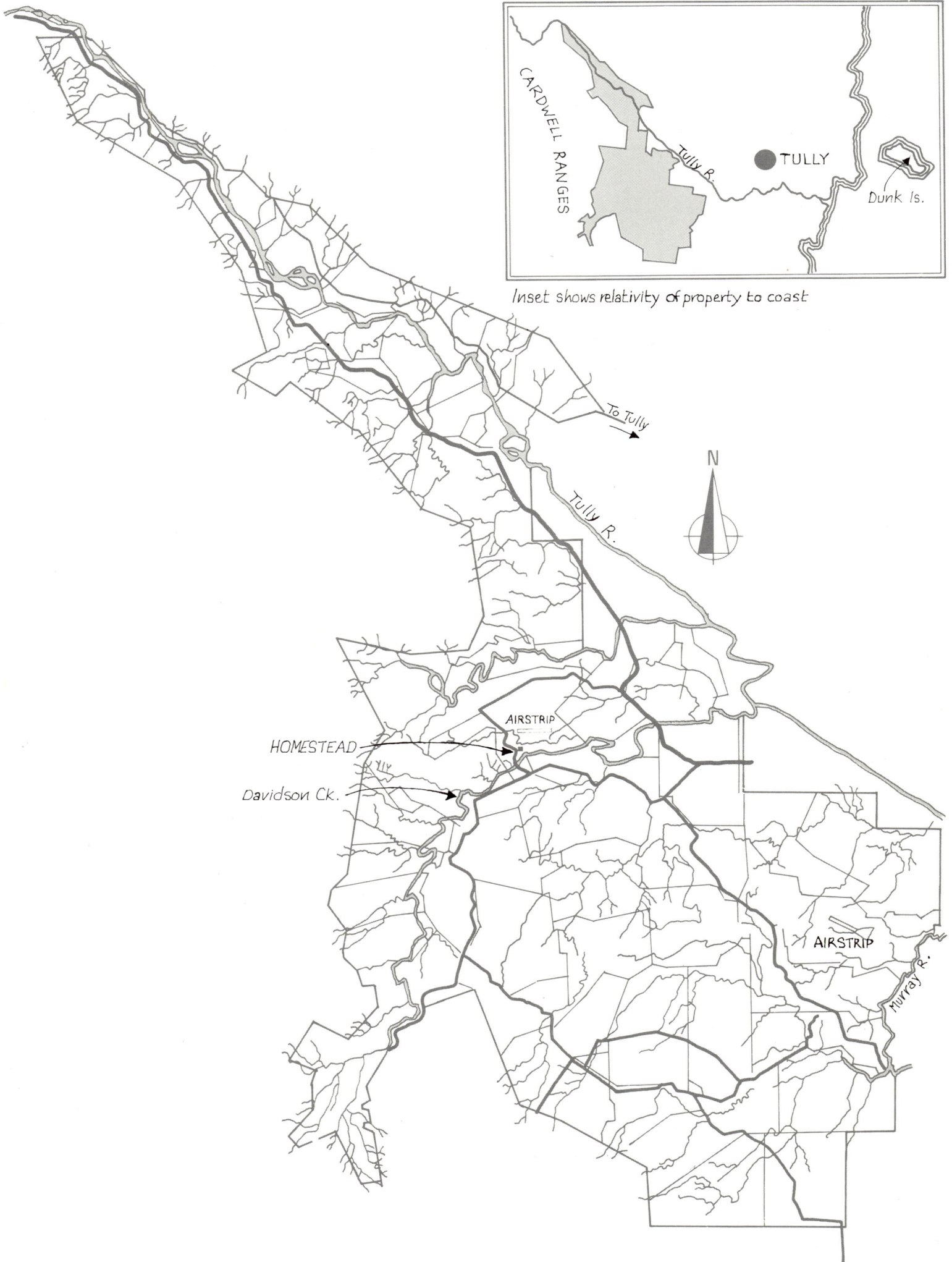

CARDWELL RANGES

Tully R.

● TULLY

Dunk Is.

Inset shows relativity of property to coast

To Tully

Tully R.

N

AIRSTRIP

HOMESTEAD

Davidson Ck.

AIRSTRIP

Murray R.

tourists on Dunk Island look across the water to the fresh beauty of the Cardwell Ranges on the mainland. They are the same mountains that dominate this station. Their covering of lush rainforest on the one hand, and the blue of the Coral Sea on the other, make this an unlikely place for a cattle station. Indeed, until recently it was an unlikely place for almost anything.

Edmund Kennedy landed here in 1848 at the start of an expedition to Cape York that was to cost him his life. South of Tully River he tried to find a way through the swamps and rainforest and to cross these mountains that stood like a barrier ahead. It was three weeks before his party could leave their camp on the beach and nearly a week after that before they could attempt to cross the mountains. They left their carts behind, but even so one of them wrote, 'We travelled up the hills all day, but made very little progress, owing to the great labour of clearing.' It was a labour that was to deter men for more than a hundred years.

When this area was later settled it was the accessible coastal plain that was opened up and used for sugar farms. The swamps and rainforest-covered mountains remained untouched, and for the most part they were as wild and impenetrable as they were when Kennedy was there. The Queensland government offered favourable terms to anybody who would clear the land and make it productive but there were no takers. Until 1963.

Then Bob Kleberg, Chairman of King Ranch in Texas, visited Tully with the Chairman of his Australian subsidiary. They spent weeks exploring the country on foot and on horseback and even then they were unable to penetrate all of it. They saw rugged mountains, untouched jungle, dank swampland and heavily timbered forest, all worthless for grazing cattle.

But Kleberg had experience of developing similar country in South America and he could see potential where others saw only ruin. He said later, 'I don't believe any ranchman in Australia had any idea what this rainforest area could be turned into.' He was right, and most Australian 'ranchmen' thought he was no more than a fool who would soon be parted from his money, considerable though it was.

Kleberg obtained a lease from the Queensland government and later that year brought in L. F. Wilkinson from Texas to start the clearing. It was a daunting task and at first progress was painfully slow, hindered as they were by 'snakes, flies and any animal that was able to sting, spit, bite or otherwise abuse you'. Soon, though, they had opened up enough country to bring in heavy machinery and then the pace quickened. Before long they had five bulldozers on the job and they were putting in nearly 800 hours each month.

Five years and more than five million dollars later King Ranch had transformed this land into a cattle station. It was, as Kleberg said, 'a tropical grass wonderland, as beautiful as it was productive', and the vision he had in 1963 was now a reality.

As a cattle station, it had very real advantages. The rainfall in the Tully district is the highest in Australia and on the property it varies from about 80 inches a year in the southern end to a staggering 180 inches in the mountains in the west. Most of it falls between January and March and there is generally a dry period from August to October. The average maximum temperature is 28°C and there is no frost.

The result of this tropical climate is that growth is never dormant. Pastures are lush and green all year round and in good seasons the paddocks can produce more grass than the stock can use. It can be three metres high and it is impossible to drive through it without having a man standing on the back of the ute to call out directions. They put a tractor in to cut it down and then turn cattle on to it and what they do not eat is returned as goodness to the soil.

When much of eastern Australia was in the grip of drought, and elsewhere sheep and cattle foraged as best they could in brown and dusty paddocks, Tully River remained rich and lush and its cattle grew fat.

Indeed, fattening cattle is the purpose of Tully River. Although there is a small Santa Gertrudis stud, the property was developed with the intention of receiving 'store' bullocks from other properties and fattening them on these lush paddocks for eventual sale to the meatworks. In particular it was intended that eventually most of the bullocks would come from other King Ranch properties which were unable to fatten them as quickly on their own, harsher, country. That principle still holds, but the carrying capacity of Tully River is so high that it can run far more cattle than can be supplied by those properties. The result is that they now supply about 20 per cent and the rest are bought in from other, unrelated, properties in Queensland.

Once the land was cleared, cattle numbers grew rapidly. From none at all in 1963, this 52,000-acre property now runs about 22,000 head of cattle and turns off about 10,000 to the meatworks every year. The stocking rate seems alarmingly high, but it is actually well below its potential. Prudence and market conditions apply the limitations, not the country.

On the face of it the economics are relatively simple. It costs about $75 a year to keep one beast in board and lodging. Each beast will stay on the property for twelve to eighteen months. During that time it must gain enough weight to recover at least that cost when it is sold to the meatworks. If its value increases by $75–100 then it will have paid its way. More than that and it will have made a profit.

But economics are rarely simple in the pastoral industry, and this is no exception. If the price of beef falls, then even high-grade cattle might not cover their costs. In those circumstances it would also be possible to buy store cattle very cheaply, but the profit on them would not materialise for eighteen months, and only then if the price at the works improved.

In some respects this is true of any kind of cattle-raising — profit ultimately depends on the market price at the time of the sale. But at Tully River the operation is so intense and costly that the difference between success and failure might be no more than a fraction of a cent per kilo.

With knife-edge economics, much of the success depends on selecting the best cattle in the first place. The buying for Tully River is done by Bryce Sturtridge, livestock manager for King Ranch, and he buys mostly Brahman or Brahman crosses. Some are bought from local saleyards, others are bought 'in the paddock' from other properties.

Cattle arrive at Tully River by road or rail. There is a rail siding about 10 kilometres from the property and a full train of livestock is given priority over other traffic during the night. Consequently many of these trains arrive in the early hours of the morning, but they have to be immediately unloaded and the cattle counted into the yards before being put in a nearby paddock which is kept for that purpose.

The cattle are kept there for a day before being moved on to the southern end of the property, which is used for all new arrivals. They stay there for three or four months so that they can recover from their journey (and often the harsh conditions they have left behind) before being moved to the richer pastures further north.

With dawn now turning into day, Richard Luck leaves the homestead high on the hill and drives to the station area below. Born in 1945, he grew up on Portland Downs, a sheep property near Longreach in Queensland which his father managed for thirty-two years. He was educated there by Primary Correspondence School before moving to Rockhampton Grammar School.

When he left in 1961 he started work at Portland Downs as a jackeroo but, finding that he preferred cattle to sheep, he joined CSIRO as a jackeroo at Belmont Research Station. There he was involved in the development of the Belmont Red, a new breed successfully developed by CSIRO for use in tropical areas. He left in 1965 to experiment with city life but it was not for him and the following year he joined King Ranch as a jackeroo at Elgin Downs under John Cooper.

He married in 1968, by which time he was head stockman, and in 1979 moved to Tully River as assistant to Neil Alderman, the first manager of the property. When Alderman left in 1980 Richard Luck was appointed manager and moved from a cottage in the low country to the impressive homestead a few kilometres away.

There was much to do. Although the property was in good shape he thought the first phase had now been completed and that it was time to move on to the next. Not that it needed dramatic changes, for the original concept was as valid now as when Kleberg had first planned it. It was more a matter of tightening up the property, toning it so that the strengths could be strengthened. Now, as he arrives at the maintenance area for the start of another day, he is a confident and experienced manager.

In the big workshop Brian Hood, the machinery foreman, is already at work on a tractor. They have recently bought a new tractor and this one will be sent to another King Ranch property in western Queensland. The workshop facilities there are not as good as they are at Tully River, so they are giving it a major overhaul before it leaves.

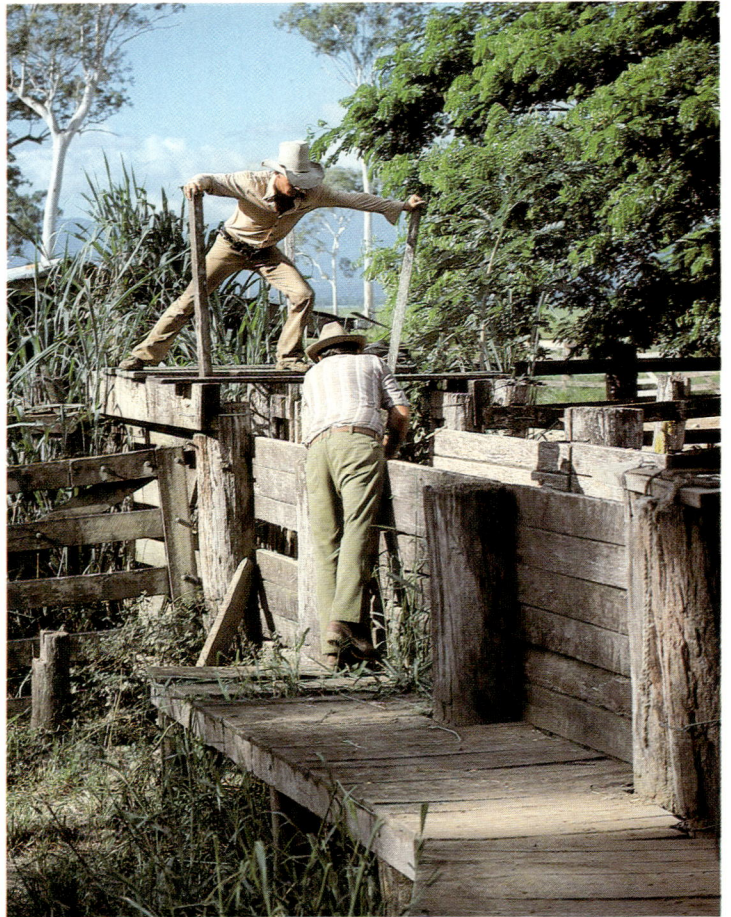
Working the drafting gates from a platform above the race.

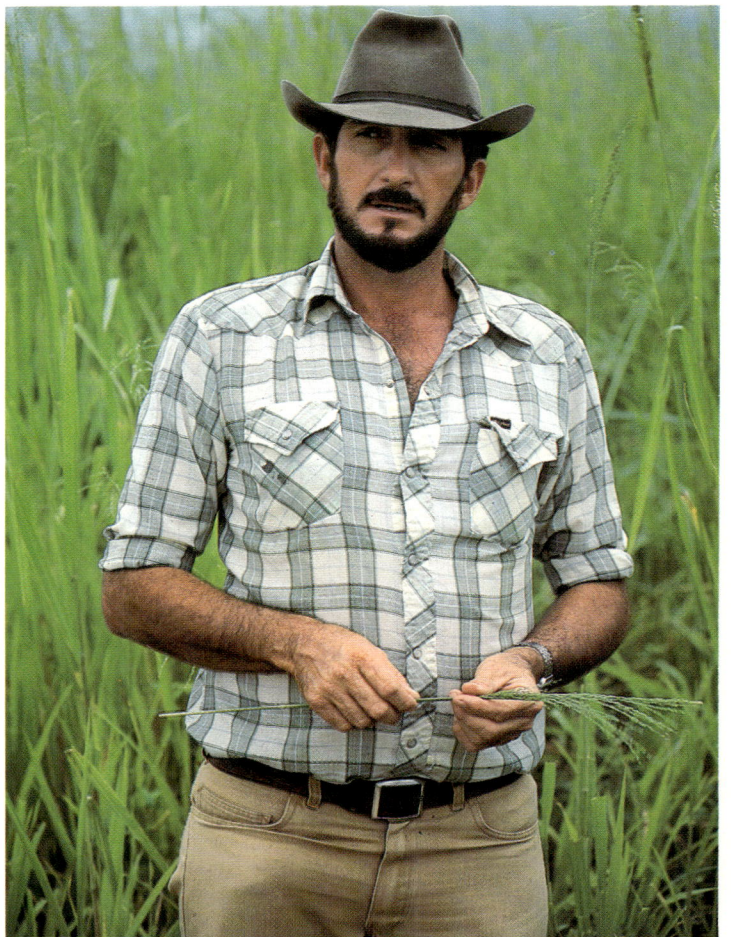
Richard Luck, manager of Tully River.

Station quarter horses in a lush paddock at Tully River. Parts of this property
receive 180 inches of rain each year.

Assembling cattle at the head of the race.

At a few minutes to seven a ute swings into the yard with half a dozen men standing on the back. They climb down and each swings a haversack and a water bottle on to the back of another vehicle. The men make up one of the tractor camps and they banter each other as they wait for the drive to the paddock. It is Friday and spirits are high. Because Tully River has none of the remoteness of an outback cattle station most single men leave the property for a weekend in Cairns or Townsville before returning for the start of work on Monday.

There are two stock-camps on Tully River, although the term is used differently here. Normally it means a team of men under a head stockman which camps out in the further reaches of the property to work the cattle. Often the men are out for weeks on end and even when they return to the station area it might be for no more than a day or two. But Tully River, although big for a fattening property, is much smaller than many stations and men can reach any part of it after a fairly short drive. So the men work as a stock-camp, but they work from home. One camp works the southern part of the property, whilst further north the senior head stockman, Bob Youngman, runs the other camp from a small outstation at Stony Creek.

Richard Luck is soon busy talking to Brian Hood in the workshop. Most stock work can be planned days or even weeks ahead, but maintenance work is not that predictable. Although some of it is routine, much of the time is spent doing unexpected repairs to machinery in various parts of the property. He is still making plans when the tractor camp leaves for the paddock, dust swirling from the vehicle as it turns on to the road.

Tractor camps are not a common feature on cattle properties but they perform a vital role at Tully River. Clearing the ground in the Sixties was not the end of the story. With a tropical climate promoting prolific growth, the country would revert to forest in only a few years if it were left alone. So clearing has to go on all the time to keep new growth at bay. Nor is it only regrowth, for clearing the land opened it to a whole range of weeds, shrubs and other plants that had not been there before. One of these weeds, called giant sensitive, is a thorny bush that curls its leaves when they are touched. Cattle will not eat it and, more importantly, it can spread at an alarming rate and smother good grass in the process. Spraying it is a full-time job and costs thousands of dollars each year.

Whilst much of this work has been going on ever since the land was cleared, an increasing amount of time is now spent on rejuvenating the paddocks. They are cleared again, ploughed and then resown with grass, frequently of a different variety. Because Tully River was the first large-scale attempt in Australia to run cattle in a tropical rainforest area there is not much expertise to draw on.

Most cattle stations are in areas where there are genera-tions of accumulated experience available, but at Tully River they have to find their own solutions. Elsewhere on the property the Department of Primary Industry is carrying out

Spraying weeds — a constant job at Tully River.

tests to determine the best pasture grasses, the best fertilisers and the best preparation and care. As knowledge and experience grow, much of the educated guesswork will disappear. But at present they are still facing new situations and precise answers are rarely available.

At 7.30 Richard Luck drives out of the maintenance area and waves to a girl sitting in a bus shelter near the road. She is going to school in Tully and has put her bike on the stand nearby. She is early, but it is a good time to catch up on some reading.

With families, there are about sixty-five people living on Tully, of which about twenty-nine are employed on the station. It is a large number for a property of 52,000 acres and is an indication of the amount of work needed for this type of operation, especially as less than half of them actually work the stock.

Because the station is so close to Tully and is surrounded by sugar farms, many members of families are able to find work nearby, although much of it is seasonal. They can also enjoy the beautiful beaches which attract thousands of tourists each year but which, being little more than half an hour's drive away, they take for granted.

and a huge bush whose flowers change from white to pink during the day. Beyond are the luxurious guest's quarters, a self-contained building containing three bedrooms as well as a kitchen and lounge.

As Richard Luck parks the ute he barely notices the magnificent view that the homestead commands and instead hurries into his small office which has no view at all. He tries to spend some time here every morning because on a property such as this good planning is essential. He has over a hundred paddocks on the station and making sure the right mob is in the right place can be a tricky business. He also has to make sure that cattle are being turned off to the meatworks at regular intervals and that they are moved to the yards for loading when the time comes.

His relationship with the meatworks is crucial. As cattle become ready for turning off he has to make sure the works can take them. Usually they can, but if not the cattle might have to stay on the property for a few more days and occupy a paddock that had been earmarked for another mob. On the other hand, if the meatworks lands a big overseas order, or if cattle from another property are delayed, then they might ask him to send cattle earlier than he had planned, which leads to more juggling.

But nobody can run a cattle station from behind a desk, or would want to, and after about an hour he is off again, this time to a set of yards in the southern part of the property where the men have assembled a mob for drafting. By the time he gets there the head stockman, Robert Braes, has already started drafting and Richard Luck climbs on to the timber platform beside the drafting race.

It is hot now and the mob of cattle raises clouds of dust as the men move some into a smaller yard at the head of the race. At the other end a man stands on a platform above the race where he can open or close the side gates to send the beast into the appropriate yard, or allow it to go straight on and through the dip.

The mob has been on this part of the property for three months and they are ready to be moved to the even lusher pastures further north. Or at least some of them are. Some are not quite ready and need another month or two before being moved. Others are judged to be as fat as they are ever likely to be and they can now be sold to the meatworks. They are not necessarily the best cattle, it is simply that their build and weight means that they would show little improvement no matter how long they were kept. Others are selected for sale to local butchers, a seasonal business but useful nevertheless, whilst those with more potential are selected for the move to the north.

It needs a good eye to select them, and it needs to be done quickly so that as each beast comes down the race an instruction can be given to the man controlling the gates. But if a beast is difficult to judge it is held in the race whilst Robert Braes looks at it in silence for as long as he needs. It is not a speed contest and it is much better to have the correct decision rather than a fast one.

Whilst it might seem nearly idyllic, especially for a cattle station, the weather can be unpleasant. Dramatic thunderstorms can be frequent and awe inspiring, with lightning flashing in vivid streaks as the thunder rolls round the enclosing mountains. The lashing rain can reduce visibility to almost nothing and if it goes on for long it can flood the road to Tully.

Richard Luck drives back up the hill to the homestead, listening to the two-way radio on the way. One of the men is going into Tully and a mechanic is giving him the number of a spare part to pick up whilst he is there. For Richard, the radio is an essential tool and is probably worth an extra man. It allows him to run the property from wherever he might be and without the help of an assistant manager. There are certainly times when an assistant would be valuable because he can be in only one place at a time, but careful planning makes those situations rare.

The homestead is a large timber house with the living accommodation on the upper floor. Below is a garage, a spare bedroom and a couple of offices. The garden is neat and lush. Bougainvillaea grows in colourful profusion and further down the slope are date palms, lychee trees, ginger plants

Richard looks at the cattle that have already been drafted and discusses them with Robert. He is like any other experienced manager in any business: tempted to do the job himself but knowing he has a responsibility to train his staff and help them acquire their own experience. It is not easy, especially as the decisions being made here will have a direct bearing on the profits of the station.

Although no two men would ever draft a mob of cattle in exactly the same way, he is satisfied with what he sees and he leaves the yards and climbs back into the ute. It is better to let men get on with the job than to stand over them and make them nervous. In any case he will have a look at the cattle at the end of the day when the drafting is finished. Then he might run some through again, but he doesn't do that very often.

He goes back to the homestead for a quick meal and looks at the mail. Then he is off again, this time to look at the paddocks in the far north of the property and which border on the uncleared rainforest.

As he splashes the ute across the Tully River his wife Gaye calls him on the radio.

'Richard, somebody just phoned to say they've seen cattle loose on the Cardstone Road near the cutting.'

'OK. I'm not far from the road — I'll have a look. Reecie, make your way over there as well, will you?'

'OK, will do.'

He drives through a couple of paddocks and then turns on to the public road which leads to Tully. He drives quickly along the bitumen and after about a kilometre he sees a bullock at the side of the road. He doesn't stop because he knows there will be more than one somewhere. As he passes the bullock it pulls its head back into the long grass and disappears.

A few minutes later, surprised at not having seen the rest, he stops and gets out to look in the grass at the side of the road. As he does so Ron Reece pulls up behind him and says he saw four bullocks further back, close to the one that Richard saw. Reecie's mate is sent to open a gate into a nearby paddock and Reecie walks into the long grass as Richard gets back into the ute. The radio calls him again.

'Richard, John Cooper is on the phone from Elgin Downs about the cattle they are sending from Avon. He says...'

At that moment twelve bullocks burst out from the grass and charge almost noiselessly across the road, followed by Reecie in hot pursuit. Richard drops the microphone and races into the long grass after the bullocks and he and Reecie flush them back to the road. Reecie tails them on foot as Richard returns to the vehicle and eases it forward to follow them.

'Richard, are you there?'

'I am now.'

'They want to know if you can freight them from Townsville if they...'

A bullock bursts from the small mob and runs back the other way, side-stepping Reecie with the ease of a winger. Richard swings the ute to head it off, but it evades him just as easily.

'Find out when I can phone them back, will you? I'm chasing bloody cattle here...'

He guns the ute and chases the bullock down the road. As he draws alongside it he turns the ute like a stockhorse and the bullock stops and looks at him. Then, as if knowing that the game is over, it turns round and trots quietly back to join the others.

By now Reecie's mate has them blocked near the gate, waiting for Reecie and Richard to help him put them through. But when they try, the bullocks break and head off in different directions. The three men run in front of them, waving their hats and whistling to make them turn back. They do so and now trot sedately through the gate without a care in the world. That leaves only the four breakaways Reecie saw earlier.

'Richard, you can call them up to 4.30.' He looks at his watch. 3.15.

Two kilometres up the road he sees the four bullocks trotting towards him with a truck driving slowly behind them. Richard drives to face the bullocks and then reverses in front of them until he can stop. He gets out and the bullocks head for the cover of the grass. He drives them back on to the road and waves at the truck as it goes past. Then he heads them up the road to where Reecie is waiting. Soon they are in the paddock with the rest.

Richard gets back in the ute and leaves Reecie and his mate to find the break in the fence. When cattle are put into a new paddock the more experienced beasts will walk round the boundary testing the fence in the hope of finding a weakness. It need not be much — one loose post is enough — but now it has to be found and repaired.

He looks at his watch as he starts to drive back to the homestead and wonders why things like that always seem to happen on Fridays. 4.10. He might just make it.

The radio again. A maintenance man is calling to ask if he knows the location of some underground pipes that need replacing.

'No idea. They were put in by a contractor so you will just have to poke around.'

'Yeah, OK — I thought there might be an easier way.'

Richard looks at his watch again.

'There never is an easy way.'

THE PILOTS

It is not surprising that aircraft were in common use in the vastness of northern Australia long before they became a familiar sight in more settled areas. With small communities separated by huge distances, the benefits were more obvious there than they were in other parts of the country. Indeed, the Australian national carrier, Qantas, was founded in 1920 as Queensland and Northern Territory Aerial Services and originally operated from the small Queensland town of Longreach.

As air services were developed many cattle stations were used as airports on domestic routes and at least one, Brunette Downs, was used as a refuelling stop by international flights. It is only comparatively recently, however, that stations have bought aircraft for their own use. Now, many of them own helicopters or fixed-wing aircraft which are flown by pilots employed by the station, whilst others hire aircraft and pilots from specialist charter companies.

Aircraft are used for a wide range of jobs. Perhaps the most spectacular is the mustering of cattle, but they usually do much more than that. They are used to check fences and bores, to take supplies and mail to distant outstations, to check that a paddock has been mustered of every beast (particularly important in disease eradication), and to fly staff to and from the nearest town, often hundreds of kilometres away, or to other properties.

The type of aircraft used depends largely on the work it will have to do. For mustering, it is difficult to beat the helicopter. It can land almost anywhere, it is highly manoeuvrable, and its ability to hover is of enormous value. Its disadvantage is that it is slow when flying across country and this makes a fixed-wing aircraft preferable when there is much 'ferrying' to do. Fixed-wing aircraft are used for mustering as well but they are not as manoeuvrable as a helicopter and have the added disadvantage of having to maintain a minimum speed in order to remain in the air.

Brian Mansfield is a helicopter pilot employed by a charter company in Darwin which specialises in aerial mustering. He was born in England in 1947 and emigrated to New Zealand with his family when he was seventeen. He started work as a motor mechanic and at the same time took his private pilot's licence on fixed-

Brian Mansfield, helicopter pilot. He works for a charter firm in Darwin and is an expert in aerial mustering.

wing aircraft to help in deer hunting in the mountains of the South Island.

From there he moved to Sydney and divided his time between working as a mechanic and towing gliders from the nearby airfield at Camden. Realising that he would never be able to afford his own aeroplane, he decided to use the money he had saved to take out a helicopter licence instead, and returned to New Zealand to do this.

He has been flying helicopters ever since and is now an expert in aerial mustering. 'In some respects it is like mustering with a horse, but it also has differences. You can climb high enough to see the whole of the mob and that means you can see leads developing much more easily than the stockmen on the ground. But you still

Bernie Regan, chief pilot of the Stanbroke Pastoral Company, beside the executive Beechcraft.

need to understand the psychology of cattle, especially if you are working them by yourself.'

After many hours of professional flying he has had only one accident. 'I had a mechanical failure in New Zealand and the helicopter fell on to the side of a mountain. I scrambled out, then watched the machine roll 200 metres down the hill.'

Bernie Regan is a different kind of pilot. Born at Bellingen in New South Wales in 1944, he joined Stanbroke Pastoral Company as a fixed-wing mustering pilot when he was thirty. Now the company's chief pilot,

he commands its privately owned fleet of seven aircraft and is responsible for selecting, training and organising the company's six full-time pilots who are based on the company's properties.

The company owns eighteen cattle stations and Bernie Regan spends much of his time flying senior executives on inspection tours in a Beechcraft. 'Although it is expensive,' he says, 'the benefits are undeniable. We simply could not hope to be as efficient any other way.'

This is true of much of the flying which is now a daily routine on many cattle stations in northern Australia.

BRUNETTE DOWNS

You are unlikely to see many people round the homestead area of a large cattle station during the day. They will be either out in the paddocks, working in the maintenance area, or looking after administration in the office. You might see a few people at the end of the day as they return home to clean up for the evening, but even then this area will not be crowded. A ute might drive in from the paddocks with half a dozen stockmen, but they disperse quickly in the fading light and silence takes over once more.

It is no different at Brunette Downs. People finish work and walk across the open space of the station area alone or in small groups, but only for a few minutes. Even on a station as big as this it does not take long to walk between buildings. Later, though, you might see more people walking to a small detached building not far from the station store. It used to be a saddle shop. Now it is the Saddler's Arms and it is the licensed pub at Brunette Downs.

When Australian cattlemen talk of men they know and properties they have worked (and in lonely stock-camps or country pubs it is a popular subject), it will not be long before they talk of Brunette Downs. For somehow this huge property in the Northern Territory is the stuff of legends — all based on incidents and people which, like the property itself, seem larger than life.

Now owned by the Australian Agricultural Company, Brunette Downs covers 12,250 square kilometres, has 4,800 kilometres of roads and 2,200 kilometres of fencing, and keeps Eastern time — 'Brunette Downs Time' — instead of Central time. It is not the biggest cattle property in Australia, nor is it the oldest. It is simply Brunette Downs and it is, as cattlemen say, 'on its own'.

If they are talking about Brunette Downs in a pub in Mount Isa, its nearest service town, they will tell you how to get there with an understatement that characterises these men. Drive west out of town, they will say, and take the first bitumen road on the right.

By the time you reach this junction you will be more than 450 kilometres from Mount Isa and well inside the Northern Territory. And as you turn on to this road, called the Tableland Highway and one of the major beef roads built by the government to aid the cattle industry in northern Australia, you will still have 140 kilometres to go.

To some it is a lonely road, for there are no buildings of any kind and not many fences. But it is not so to the cattleman, for he is simply driving through paddocks, big though most of them are. He will recognise the southern boundary of Brunette Downs, featureless though it might be to others, and know that he is now driving through Boree paddock. Further on he will pass through White Hole paddock and then cross Brunette Creek before turning left on to a dirt road that leads to the homestead area. By the time he reaches it he will have driven 50 kilometres from the southern boundary and if it is the end of the day he will be looking forward to a beer at the Saddler's Arms.

It was in 1883 that this land was first taken up. Most of western Queensland was at that time badly affected by drought and John and Tom MacAnsh agreed that it was time they moved their herd west from Albilbah Station in search of better grass and perhaps to start a new station beyond the Queensland border.

It was not an easy journey. John MacAnsh broke a leg at the Diamantina River near Boulia and had to be sent by dray north to Burketown, where he could be put on a ship for Brisbane. Tom, who had not long left school, decided to go on alone. But by now the drought had caught up with him and he had to hold the mob at the waterholes around Boulia until the rain came and made it safe to take the cattle over the little-known country to the west.

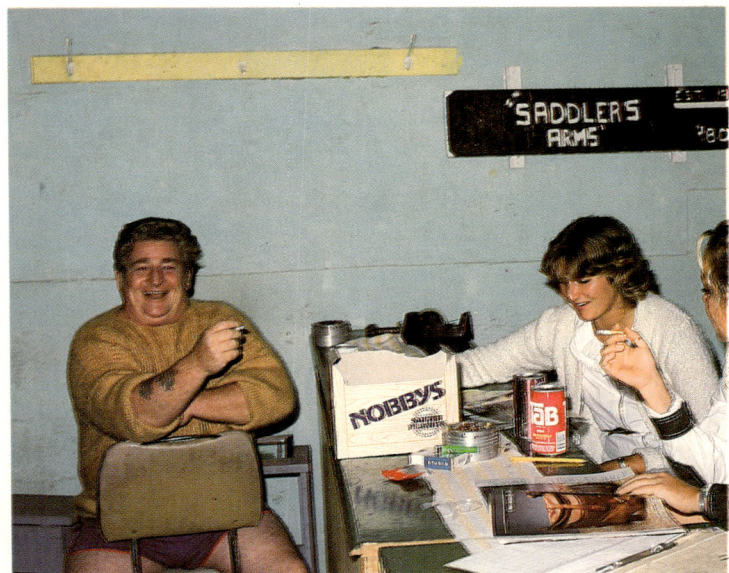

The Saddler's Arms, the licensed pub at Brunette Downs.

Although the lack of feed and water meant that the cattle could not be moved with safety, men and horses were not so restricted. So, with his cattle held for the duration, Tom MacAnsh decided it was a good time to go ahead and find 'a place'. He needed a well-travelled stockman who was also an experienced bushman, someone who knew the country ahead and who would also be able to move the mob up when the time came. The man he found was Harry Redford. Known throughout Australia, his qualifications were impeccable by almost any standard.

A little over twelve years earlier, in March 1870, Redford and four men had illegally mustered over a thousand head of cattle in the remote country of Bowen Downs Station near Aramac in Queensland. Then with two of the men he drove the cattle off the property and slowly moved them south-west. It was a good season and with no shortage of feed and water he was able to move with relative ease. Through the

Channel country of south-west Queensland, he took them down Cooper Creek, where the Burke and Wills expedition had met its fate only a few years earlier, and then down the Strzelecki Creek. It was epic droving, as outstanding as it was illegal.

With the Wet over, the overseer at Bowen Downs discovered the tracks near the lonely yards and he was instructed to take a man and follow them in the hope of finding the huge mob of missing cattle. When he reached Walleldine Station in South Australia he found a distinctive white bull which Redford had sold there, and as he journeyed further south he found small mobs of cattle which Redford had sold on the way.

Redford was caught and eventually stood trial at Roma in southern Queensland. By this time his adventure had caught the imagination of the Outback and, against all the evidence, the jury brought a verdict of not guilty. The judge was furious. 'I thank God, gentlemen, that this verdict is yours, not mine,' he said. But by that time Redford was being carried from the court by a cheering crowd.

Now, Tom MacAnsh and Redford journeyed across the Barkly Tableland and together selected the land that was to become Brunette Downs. When the drought broke, Redford returned to Boulia and brought the mob of 4,000 head of cattle on to the new property.

Redford managed Brunette Downs for a few years before taking up a run of his own at Corella Creek some 50 kilometres away. He was not successful, however, and he soon sold it to MacAnsh and went as manager of McArthur River Station further north. He stayed there for many years and became recognised as one of the best cattlemen in the Territory. In the end, though, he was 'allowed to leave' by the owner's son who, having been trained by Redford, now thought he could run the property on his own.

Redford returned to Brunette Downs as a stockman but it was not a happy situation, especially as the manager there had originally been taken on as a stockman by Redford. He was 'allowed to leave' again and on 21 March 1901 Redford, now fifty-nine, left to ride to Tennant Creek. But Corella Creek, on the part of the property he had once owned, was in flood and Redford drowned as he tried to cross it. His body was recovered by a native woman and buried nearby.

In 1904 Brunette Downs was taken up by two brothers, James and Francis White, who were members of a large family of property owners descended from James White, who had arrived from Britain in 1826.

Because of commitments to the other family properties they ran Brunette Downs as absentee owners, but in 1912 they were joined by A. J. Cotton and he made regular visits there. By the time James White died in 1926 his son, also called James, and his nephew, Harold Fletcher White, were already doing much of the work. They bought Cotton out of the business and after a series of purchases from other members of the family, in 1932 James White obtained control of Brunette Downs as managing director of the Gulf Cattle Company.

He was determined to make the property even more successful than it had been in the past. In 1934 he appointed the legendary Tom Barnes as manager and with head stockman George Lewis they started a massive programme of improvements. Step by step, and over a period of years, they put in over 3,000 kilometres of fencing, sank 30 bores, built 30 sets of yards, cut hundreds of kilometres of roads and upgraded the herd of Shorthorns. In the process Brunette Downs acquired its reputation as a trend-setter and became one of the most famous cattle stations on the continent.

But they could not control the seasons. In 1951, six years after the death of James White, the property was in dreadful drought once more. James's son, James S. White who had joined Brunette Downs as a jackeroo in 1934 and had since taken over from his father, was faced with a huge problem.

He needed to reduce the stock urgently to give those remaining a chance to survive, but the only way to remove

The station road train.

stock at that time was to walk them out. The problem was that with no feed along the stock routes there was almost no chance of them surviving the trip. In the end he bullied a politician in Brisbane to guarantee transport from the railhead at Dajarra, some 650 kilometres away, then dropped feed every 60 kilometres along the stock route before sending off 12,000 steers in a number of mobs. The droving was accomplished without the loss of a single beast. But they were not so fortunate on the property, where they lost over 30,000 head.

In 1953, with the drought over, Brunette Downs was the site of one of the most unusual cattle auctions the industry has ever seen. Organised by the Shorthorn Society of Australia, thirty pedigree bulls were gathered in Melbourne and flown in Bristol freighters in what was said to be the biggest airlift of cattle ever attempted. Over two thousand spectators and buyers came from all over the world and competed fiercely for some of the best Shorthorns that Australia had to offer.

But for the White family it was almost the end of the road. They had taken an undeveloped property and over the years had turned it into one of the most famous in the world. They had ridden out droughts and floods, good times and bad, but there was one problem that even they could not overcome: old age. By now most of the directors were in their seventies or older and James S. White was the only one who could exercise any degree of supervision. After much thought, and with much regret after owning it for more than fifty years, in 1958 the White family decided to sell Brunette Downs.

In spite of its now considerable reputation, Brunette Downs was not an attractive proposition. It was in drought once again and apart from the immediate difficulties there seemed no way of preventing the same difficulties arising again in future droughts. The records showed how unpredictable the rainfall could be and buyers knew that they were likely to face the same crippling situation at some time in the future, and perhaps a not-very-distant future at that. In an industry that survives by taking risks, this seemed too high for most.

Bob Kleberg, then Chairman of King Ranch of Texas, could see the problem. With the rivers and creeks long since dry, the stock had to depend on the forty-six bores for their survival. But on a property as big as this the bores were too far apart to keep them alive. The grass had been eaten out or trampled for about 8 kilometres round each bore and these barren areas now contained thousands of dead and dying cattle. Beyond them there was still ample grass — dry to be sure, but still capable of supporting stock — but it was too far away from the water to be of any use. It meant a 16-kilometre walk there and back for the cattle and few could hope to do that for very long. Even if they could, the grass would continue to recede as it was eaten out and the walk would become even longer.

For those who could see the problem (and it was very clear from the air), the solution was obvious. The property needed far more bores to keep the stock alive in droughts such as this. But bores were expensive and few had the capital necessary to sink enough to make the property drought-proof.

But Kleberg had. In December 1958 Brunette Downs was bought by King Ranch's Australian subsidiary for $2 million. It was the start of an investment programme that was to go on for years. They spent another $2 million on water alone and by the time they had finished they had installed another 83 bores and most cattle were now no more than 3 kilometres from the nearest water.

The Shorthorn cattle, which had been raised at Brunette Downs since MacAnsh's day, were replaced by Santa Gertrudis. Originally developed in America by King Ranch for hot, dry areas, they had been built up on other King Ranch properties and had already proved extremely successful in northern Australia.

People were needed too and experts, many of them young and energetic, were brought in to plan and implement the improvements. For many it was their first experience of outback life and with Tennant Creek, the nearest town, some 350 kilometres away, they had to adjust to a completely new way of life.

It was not without humour. When the single men invited the single girls for dinner, it had to be done properly. At the appropriate time the men, dressed in suits, drove their cars

The station area at Brunette Downs. The homestead is on the extreme right, kitchen and jackeroos' quarters are in the centre, and workshops are on the left. In the distance is a street of houses alongside Brunette Creek.

202

to the girls' quarters about 50 metres away. The girls, dressed for an evening out, got in the cars and the men drove them up and down the airstrip several times before taking them back to their quarters for the best dinner they could provide.

In 1970, concerned at the prospect of the USA imposing restrictions on the import of meat from disease-infected countries, Brunette Downs started its own programme to eradicate tuberculosis and brucellosis from its herd. Three years later they built what is still the only station-owned testing laboratory in Australia and which is still run by technician Frank Shiel. In it, he installed the first operational Dynatiter machine in Australia and devised a procedure for testing for brucellosis. This is now used in government laboratories which have since been established as part of the national campaign to eradicate disease from cattle.

But Brunette Downs could not escape the extremes of weather that had been such a feature of its history. In 1961 the property received little more than six inches of rain and two years later it had less than nine inches. The new bores meant that most of the stock could survive, but ironically it was too much water that caused the biggest problems.

There are three separate lake systems on Brunette Downs: Corella Lake, fed by the creek which drowned Harry Redford; Lake Sylvester, fed by Brunette Creek which also feeds a smaller lake alongside the station area; and Lake Deburgh, which is fed by the Playford River. The country round them is flat and well timbered and in average seasons these lakes have permanent water. But the seasons in the Sixties were nearly all below average and by the early Seventies these lakes were nearly dry.

Then the seasons changed. In 1973 the property received more than thirty-three inches of rain and this was repeated again in 1974. Suddenly there was water everywhere. The lakes flooded and joined together to make an inland sea as more and more water came down the watercourses. Soon the water covered more than 3,500 square kilometres, or nearly a third of the property.

More than 4,000 head of cattle were stranded on islands in the middle of this sea and they had to be swum to safety with the help of an outboard dinghy. Men, too, were cut off. Two stockmen spent a couple of days and nights clinging to a mulga tree which they had climbed when the water started to rise. By the time they were rescued they were waist deep in water and had been joined by several snakes and countless centipedes.

In spite of the damage caused by this flood (and dead trees are a constant reminder of how extensive it was), when King Ranch sold Brunette Downs in 1979 it was a very different property to the one they had bought. But in the meantime the company had developed a new fattening property at Tully River, in the lush rainforest south of Cairns, and Brunette Downs was too far away to be a source of store cattle for fattening. Deciding then to withdraw to properties in Queensland, King Ranch sold Brunette Downs to the Australian Agricultural Company.

Swimming a mob of cattle to safety during the 1974 flood.

If King Ranch had been a relative newcomer to the Australian pastoral scene when it bought Brunette Downs, the Australian Agricultural Company certainly was not. It had been started in England in 1824 as a result of a recommendation by Commissioner Bigge that land be granted to encourage settlement in the remoter parts of New South Wales. In spite of considerable opposition from the colony, the company was granted one million acres between Port Stephens and the Manning River and later, against even greater opposition, they were given the government's coal mine at Newcastle.

The first few years of the company were turbulent and far from successful, but the increasing demand for coal and the

fences to rebuild and many bores and windmills in need of major repairs. In the first three years the company spent nearly $2 million dollars on Brunette Downs alone.

Other changes have since been made as a result of the national compaign to eradicate disease which required better facilities for handling cattle. The success of the campaign depends on a clean muster of cattle for testing from each paddock and this was almost impossible in those vast areas, even though the country is mostly flat and open. The Brunette Creek paddock alone covered 1,300 square kilometres and it was impossible to manage the cattle with the efficiency that was now required.

Many of the big paddocks have since been subdivided into smaller areas, a job that was to some extent made easier by the many bores that King Ranch had put in. Even so, it is an expensive business and it must be spread over a number of years. There are now forty paddocks on Brunette Downs and the biggest now covers 950 square kilometres — still huge by any standard — but the average is now less than half that size. There are nine main stockyards capable of working 3,000 head of cattle at a time.

There are 50,000 head of Santa Gertrudis cattle on the property and about 13,000 head are turned off each year. Of these, about 8,000 are sold to meatworks and the rest are sent as stores to the company's fattening properties in the Queensland Channel country.

The property employs forty people, of which about ten are jackeroos. Many of these are city born, others come from family properties, and some come from agricultural colleges. With families, there are about forty-five people living on Brunette Downs itself, whilst about one hundred and twenty Aborigines live on a settlement not far from the station area. Some of these work as ringers and maintenance men and some of the women are employed in the station area. The whole population of this settlement is looked after by the station's full-time nursing sister and the children go to a school on the property run by the Northern Territory government.

The stock is worked by one stock-camp consisting of thirteen men and they camp on different parts of the property from April to about the middle of October, or whenever the Wet makes further work impossible. They return to the station area about every three weeks, but they are there for only a day or two before going on to the next area to be worked.

Although work on the stock-camp is as hard and demanding as it is on cattle properties everywhere, the men have a luxury not enjoyed by most. Their kitchen–dining room is a modern caravan which provides not only good food but also refrigeration and a degree of comfort that is more than welcome after a long day in the saddle. Men sleep in swags on iron-framed beds which stand, free of any cover, near the caravan.

Although the station staff sometimes make up another stock-camp in busy periods, the fact that most of the work can be done by one camp is only made possible by the use of

later discovery of gold on its land at Peel River guaranteed its survival. The company continued to increase its pastoral holdings and by 1976, when the company's head office was transferred from London to Tamworth in New South Wales, it was a major owner of properties throughout New South Wales, Queensland and the Northern Territory. Buying Brunette Downs was simply a logical step in a process that had then been going on for more than one hundred and fifty years.

Although many necessary improvements at Brunette Downs had already been completed by King Ranch there was still much to do when the A.A. Company took over. The property had not yet recovered from the flood and there were a contract helicopter for mustering. Without it, they would almost certainly need another camp working full-time.

The station Cessna is also used for this work, and particularly for locating mobs of cattle before the muster starts. This can be a hair-raising experience for those unfamiliar with it. One Sydney journalist, taken along as a spotter, wondered how she could hope to spot anything when she had her eyes closed most of the time, and described how she broke two cigarettes before she managed to get one from the packet after she landed.

The whole of the property is run from a small office building at the station, where the manager, Brian Gough, tries to divide his time between the paperwork of a multi-million dollar business on the one hand, and his need to see the stock being worked on the other.

He was born at Cairns in northern Queensland and his father, who had been involved in mining activities in the Cape York Peninsula, shortly afterwards bought a small mixed farm. After being educated at Valentine Plains School near Biloela, Brian Gough spent most of the next five years working on the farm and other cattle properties or droving. He joined King Ranch in 1963 as overseer at Tully River and then went as head stockman to Litchfield Station in the Northern Territory in 1966.

In 1969 he spent a year as overseer at nearby Tipperary and then moved to a neighbouring station, Elizabeth Downs, on the southern side of the Daly River. Four years later he became livestock manager at Mount Bundey on the Adelaide River near Darwin and in 1977 returned to Queensland to work as a contractor at Gladstone. He joined the A.A. Company in 1981 as manager of Brunette Downs.

He believes that the future of the cattle industry in remote areas now lies with companies rather than with individual owners, and points to Brunette Downs as an example of this. It needed vast amounts of investment capital before the property could hope to ride out the inevitable run of bad seasons that occur from time to time. 'It is not that individual owners lack the skill or experience,' he says, 'for some of them clearly don't. It is simply that very few of them can command the huge resources needed for improvements.'

This morning he is catching up on the paperwork whilst waiting for a meat buyer to arrive. He will have to spend most of the day with him and this is the only time he will have at his desk. Nor is it very long, for soon there is the sound of a Cessna overhead and, by the time he goes out to meet it, it is already taxiing to a halt near the homestead.

After a round of handshaking and the inevitable conversation about the weather, they all climb back on board to fly to Bloodwood paddock in the south-east of the property, where the stock-camp is working a mob of cattle in the yards.

It is about a thirty-minute flight and when they arrive the pilot is reluctant to land on the nearby airstrip. A mob of cattle has wandered across the strip and now, in the morning light, it looks like the surface of the moon. For one passenger, it is the most uninviting airstrip in the whole of Australia.

Brian Gough, manager of Brunette Downs.

Frank Shiel in the laboratory at Brunette Downs. His method of testing for brucellosis, which he developed here, has now been adopted by government laboratories.

A mob of cattle in the south-west of the property.

Aboriginal children from the settlement near the station area.

The personal gear of a stockman and his horse at a stock camp on Brunette Downs.

Brian Gough suggests that they radio for a ute to drive along it at speed to prove that it is quite firm. As the plane makes repeated low passes over the strip, the ute peels away from the camp and drives quickly over the strip. There is little dust and by the time the vehicle reaches the far end it is going flat out. The pilot is satisfied and turns to land with at least one very silent passenger in the back.

On the ground, the ute takes them to the yards to look at the cattle. The buyer climbs over the rails and walks slowly through the mob, hand on chin, hat well forward.

'These should grade all right, Brian, but I'm not sure about those over there.'

'Yes, I agree with that.'

Then they drive off to the paddocks, past the trees which died in the flood, and out to Dookamundra. They look at more cattle and this time the cattle, being on familiar territory, look back.

By the time they get back to the yards to fly to the station the morning has gone. They have a lunch of cold beef at the homestead and then leave again, this time in the Landcruiser to look at Top Paradise paddock. It is in the north of the property and is about 90 kilometres from where they were this morning.

Meanwhile Brian's wife Shirley drives to the Brunette Downs racecourse to see what will need doing before the next meeting. It is held in June each year and has been the most important part of the local social calendar since it was first held in 1910. It is a three-day meeting with two days of racing followed by a rodeo on the final day.

People come from hundreds of kilometres for this meeting, some by plane, others by road so that they can bring their racehorses with them. The course is about 16 kilometres from the station area and people set up camps on semi-permanent sites near the course. They put a sign over each camp showing the name of their property and renew friendships with people whom they might not have seen since the meeting last year. Hospitality is unstinting and seemingly continuous, but nobody forgets that they have come for a race meeting. At least not during the day.

Now, with the running rails stretching away to the distance and the camp sites empty and silent, the racecourse looks forlorn. Shirley walks round with a notebook and makes a list of things that need attention. There will be a lot of work but in a few weeks' time the place will come to life again to stage one of the great social occasions of the Northern Territory.

The homestead at Brunette Downs.

It is all part of a normal day at Brunette Downs, although there is hardly ever a day that is normal in the accepted sense of the word. In the office, bookkeeper Howard Dryden is waiting for a call to come through on the radio telephone. It is the only telephone link with the world outside and delays of nearly an hour are not uncommon.

In the kitchen Bob Foulston, known throughout the entire Northern Territory as 'Pommie Bob', is preparing the evening meal for the single men. Pat Shiel is taking readings from the meteorological instruments grouped in an open space near the office. Soon she will send the readings by radio to a receiving centre, where they will be analysed with others to form a comprehensive picture of the weather pattern over Australia. This will be the basis of the next weather forecasts throughout the country.

And at the end of the day most of them will walk across to the Saddler's Arms for a yarn and a beer. They might talk about cattle or they might not, for their interests are wide and certainly not confined to the activities of the station.

The pub closes promptly at 6.45 p.m., Brunette Downs time. The single men go to the dining room to find out what Pommie Bob has been up to, whilst the married people walk across to the row of houses on the opposite side of the station area.

It is dark when the road train pulls in from the Bloodwood yards with its load of cattle for the meatworks. The load and vehicle are checked, then it rumbles off into the night on the long trip to Mount Isa. But at least the driver has no fear of getting lost. He merely has to drive down the Tableland Highway and turn left when he reaches the first bitumen.

UNDOOLYA

By the time tourists reach Alice Springs they have learnt something about the size of Australia in a very personal way. Even by modern jet it is a two-hour flight from Adelaide and the once-a-week direct flight from Sydney takes more than three hours. If they have come by road they will have been travelling for days and they will have seen something of the emptiness of the Australian Centre.

It is an emptiness that disturbs some people. Beautiful and harsh, its barren loneliness stretches away to distant horizons that are barely visible in the shimmering heat of the day. People can die out there, the tourists tell each other, and they know it to be true.

The pleasant town of Alice Springs comes as a relief, like an island surrounded by a limitless sea of aridity. Shops, motels, cars and supermarkets — it has all the reassuring trappings of a modern town, but it is also different enough to make tourists feel that the long journey has been worthwhile.

Dusty Land-Rovers drive in, their aerials lurching from side to side, their roofs stacked with gear, as a reminder that people do venture into this wilderness. And in the hotel bars European and Aboriginal stockmen drink together and talk of experiences shared on cattle stations not far away.

For the tourist, Alice Springs offers a sense of adventure that is not readily found elsewhere. It is the 'real' Australia — outback, isolated Australia — and suddenly they feel part of it, everyone a pioneer.

The sense of isolation is very real. Although there are a few small settlements along the single road that runs from Adelaide to Darwin linking the town with the world beyond, the nearest large towns are 1,600 kilometres away and there is not much in between except this hostile country. This isolation starts at the edge of town, but now it is a vicarious danger to be enjoyed because of its closeness and in the reassuring knowledge that it does not have to be confronted.

Intrigued by its uniqueness, some tourists wonder what this place was like when it was first settled. What sturdy men trekked their way into this heart of Australia, and why on earth did they do it? It is breathtakingly beautiful, dominated as it is by the massive MacDonnell Ranges, but there must have been a better reason than that. And there was a better reason. It was not to run cattle, as many think. It was to build a telegraph line.

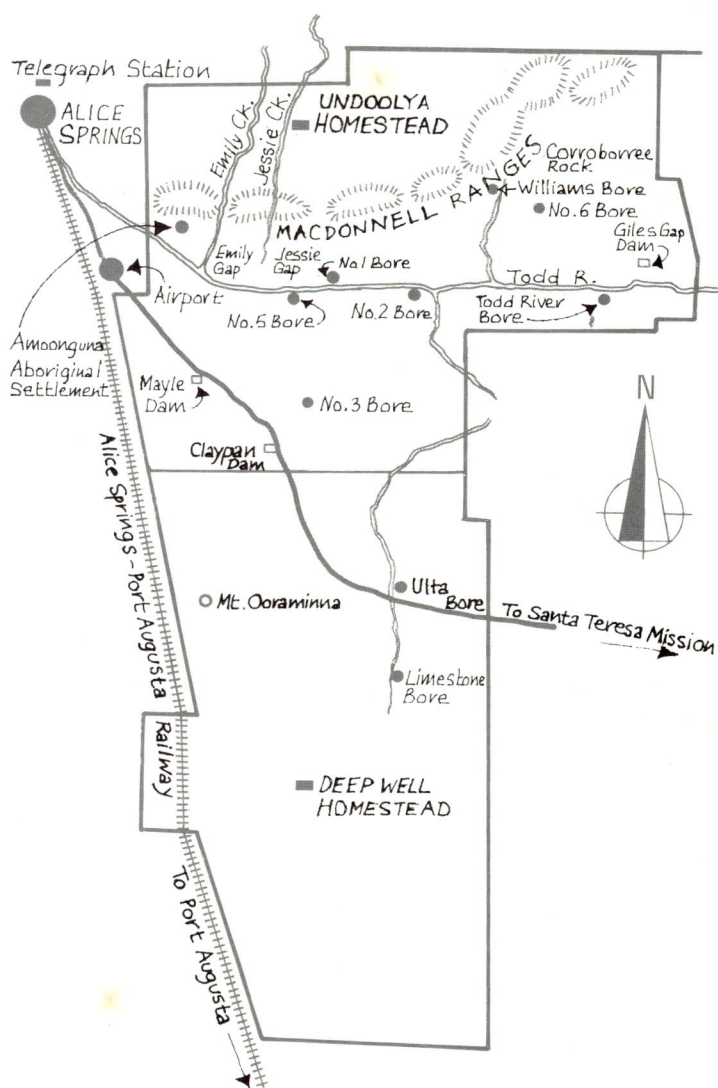

It was a preposterous idea. In 1870 the South Australian government undertook to build a telegraph line from Adelaide to Darwin to link up with the international system that was then edging its way through the Far East. Spurred on by Charles Todd, the head of the South Australian Telegraph Department, they started to build this line across country that had been crossed only once and which for the most part was totally uninhabited by white settlers. They knew little of this country except that it promised to be some of the most inhospitable in the world.

Mount Undoolya, part of the MacDonnell Ranges which dominate the northern part of the property.

Early motor cars at Undoolya in the 1920s. The large building is the original homestead built by Bagot in 1872.

William Hayes, one of the outstanding pioneers of Central Australia. When he bought Undoolya in 1907 he became one of the largest landholders in the district.

By early the following year the work was going reasonably well. Men were already working on the southern section whilst others had been sent by sea to Darwin so that they could work south from there. Meanwhile more men were slowly making there way through Central Australia to start work on the middle section of the line which would eventually join the work from the south and the north. And there they hit a problem: they could not find a way through the MacDonnell Ranges which run for hundreds of kilometres across the very centre of this continent.

With men and equipment trapped south of the ranges, surveyors were sent to find a gap that would take them through to the country beyond. When some reported that they had seen a way, W. W. Mills was sent ahead to confirm that the route was practical. On 11 March 1871 he found a dry riverbed 'with numerous waterholes and springs, the principal of which is Alice Spring, which I had the honour of naming after Mrs Todd'. So Alice Todd, who had come to Australia with her husband in 1855, gave her name to a place that has since become synonymous throughout the world with the Centre of Australia.

As men now passed through on their way north, some remained at Alice Springs to build a telegraph station there. A tiny outpost of advanced technology surrounded by nothing, it sat like a fortress beside the Alice Spring.

Reassured by its presence, before long men began to think of the unlimited land that surrounded it. They knew almost nothing about it, but it was there and that was all that mattered. In February 1872, barely six months after that part of the line had been finished, Edward Bagot applied for a pastoral lease for a block of land that started at the fence of the telegraph station and extended far beyond the MacDonnell Ranges.

Bagot had been responsible for building the southern section of the line and as the work extended further north he had presumably followed it in search of an opportunity such as this. Having been granted his lease, he called his property Undoolya — an Aboriginal word meaning 'shadow' — and built himself a small, solid homestead that was not unlike the buildings of the telegraph station. It was, apart from those, the first building in Central Australia.

Bagot did not stay long, however, and in the next few years Undoolya passed through a number of hands. In 1876 it was acquired by John and Robert Love and Andrew Tennant and soon they were running 5,000 head of cattle under the care of their manager, Alec Ross. When he married in 1885 his wife was the first white woman to come to Alice Springs.

The property changed hands in 1891 and again in 1906, when it was taken over by Norman Richardson, a property owner from Port Augusta. In 1907 Undoolya was sold again, and this time the buyer was an elderly bushman called William Hayes.

Hayes was born in Liverpool in England and arrived in South Australia in the early 1850s at the age of twenty-one. He worked on a number of properties in the north of South Australia and, with his wife, also did contract work with his horse team. They were, like most rural workers at that time, always on the move.

In 1884 William Hayes was given a contract by Sir Thomas Elder to build fences and dams on two of his properties in Central Australia: Mount Burrell (later called Maryvale) and Owen Springs. Using bullock and horse teams, he moved his equipment and materials on to the properties before returning to the railhead at Warrina some

The station buildings at Undoolya, from an outcrop near the Jessie River.

400 kilometres to the south. There, under another contract, he loaded his wagons with a quantity of steel telegraph poles which were to replace the original wooden ones that were now being eaten by termites. He delivered the new poles to the telegraph station at Alice Springs and then returned to Mount Burrell to start his work there.

At the same time he took up a lease of his own. Called Deep Well, it bordered Mount Burrell in the south and Undoolya in the north. The women and children looked after it whilst William Hayes and the older boys carried on with the contract work.

In the late 1890s Sir Thomas Elder, discouraged by his lack of success in the area, terminated his leases and moved his stock back to the south. He sold the stock that could not be mustered to William Hayes, who also took over the leases of Mount Burrell and Owen Springs. Now, with the purchase of Undoolya in 1907, he owned a vast tract of grazing land some 250 kilometres wide and whose southern border was 300 kilometres south of Alice Springs.

The Hayes family was typical of many others who took up land after the large pastoral companies had failed in Central Australia. The companies had found the distances to southern markets too great to be profitable and as they withdrew one by one their properties were taken over by families whose needs were less and whose willingness to work for themselves was unlimited.

They were the true pioneers of Central Australia and the hardships they faced are difficult to imagine now. When one of the Hayes' children was injured in an accident his parents put him in the horse and buggy and drove 500 kilometres south to the railhead at Oodnadatta, where they joined the train for the three-day journey to Adelaide. The child died shortly afterwards.

Some families, unable to cope with hardships such as this, sold their leases and followed the large companies to the more settled areas in the south where life was easier and cattle could be run with more certainty. Others, like the Hayes, battled on. They rode out the good with the bad until in time they acquired a tenuous security and their properties developed an air of permanence in a landscape which they could not hope to dominate.

With the death of William Hayes his son Edward took over and became manager of Maryvale, Owen Springs and Undoolya. Their first property, Deep Well, was sold and in 1930, when the family company was also sold, Edward Hayes bought Undoolya.

By this time Edward Hayes was already one of the most successful landowners in Alice Springs. He put his stock above everything and everybody and valued each individual beast. He also acquired a reputation for being hard and uncompromising — inevitable, perhaps, for someone who had built a successful empire where many had failed.

213

Edward Hayes was now being helped by his son Ted, who had been born at Alice Springs in 1914 and who had been eight years old when the family moved on to Undoolya. Governesses were brought in to give him an education on the property, but when none could be found he was sent to the small school in Alice Springs where Ida Standley taught European children in the mornings and Aboriginal children in the afternoons.

At that time the town of Alice Springs consisted of only six houses and had a population of twenty men and a few white women. A few kilometres from the telegraph station, which was still in use, the town had originally been given the name of Stuart. It was not a popular name, however, and soon both the town and the telegraph station were known as Alice Springs.

When Ted Hayes left school in 1927 at the age of thirteen it was with an education that was remarkably good. 'I never thought about the isolation,' he says. 'Most people in Alice Springs at that time had come from somewhere else and we learnt about the rest of the world by listening to them.'

Nor was there any doubt about what he wanted to do now. 'When you are born on a cattle station you just want to leave school so you can join in the mustering. I never wanted to do anything but run Undoolya.'

He needed experience to do that, however. So after working with his father on Undoolya for a few years, in 1935 he joined six men under a boss drover and helped take a mob of 1,200 bullocks from Elsey Station to Brunette Downs. Two years later he married a girl from the neighbouring property of Loves Creek and he and his wife moved on to Owen Springs, his father having bought the property back after it had been sold in 1930.

In 1940 Ted and Jean Hayes moved back to Undoolya and he continued to work there for wages until 1947. By then he had saved enough to buy a share of the property from his parents and in 1953, when they were ready to retire, he was able to buy the remainder. So at thirty-nine he became the third generation of the Hayes family to own Undoolya. In 1960 he bought back Deep Well and the two properties were combined again for the first time in many years.

Today the two properties cover 2,600 square kilometres of beautiful country that is so typical of Central Australia. The MacDonnell Ranges run from east to west across Undoolya and their knife-edge ridges, steep and high above the surrounding country, make a jagged backdrop that changes colour with the passing of the day. Smaller ridges run across the property in the same direction as the Ranges. To the north there are plains between these ridges, but further south, on the sandstone, the country is more undulating.

To many tourists still trying to come to terms with this arid, threatening land, it is one of life's mysteries that it can support cattle. They see red soil instead of grass and most of the trees and bushes seem deformed and barely alive. By the empty watercourses huge red gums stand as proof that some living things can flourish here, but they can see nothing that might support cattle, let alone in quantity. But Ted Hayes

Ted Hayes, the owner of Undoolya.

Bill Hayes, manager of Deep Well.

214

Jim Hayes, left, manager of Undoolya.

A stockman rolls his swag before leaving for the stock camp.

Poll Herefords at a temporary waterhole.

can. Indeed, for the past few years they have had a run of good seasons and he cannot remember the country looking better than it does now.

Even so, there is no permanent natural water anywhere on the run. The average rainfall is 10½ inches a year and most of this falls during the summer. Sometimes they have much more, and then the creeks run and the Todd River might rise to spread beyond its high banks and flood the adjacent paddocks. Usually, though, these watercourses are dry and sandy and the water that has been trapped in rocky pools along their length dries out in the high temperatures of the summer.

But if the ground is dry for much of the year, beneath it there is water in abundance. The southern part of the property is on the edge of the Merrinie Basin, a vast underground lake so big and deep that nobody knows when it might be exhausted. The property uses thirty bores to pump this water from below the ground into tanks which supply the fifty paddocks.

The harnessing of this water, even after it had been found in such great quantity, could not be done all at once. The development of the property has been spread over many years and during his lifetime Ted Hayes has seen it change from virgin land to its present abundance.

Done in stages, the first part to be improved was the area round Emily Gap. There, a bore was sunk and fences built. Then another bore was sunk further out and more fences built. As each new bore was sunk, more land was brought into use until now no bore carries more than 400 head of cattle and none of the property is overgrazed.

Light stocking has been one of the keys to success. They run 9,000 head of poll Herefords at Undoolya and Deep Well and in good seasons augment these with store cattle bought in for fattening. They turn off about 3,000 head each year and in addition sell about thirty bulls and twenty stud heifers. The light stocking rate and the increasing number of bores means that the whole of the run is good cattle country. Even the bushes, the 'top feed' which thrives in good seasons such as this, provide good feed where the tourist thought there was none.

But not all the seasons are good. When the rain failed in 1960 it was the start of a drought that was to become the worst on record. At first they thought it would last only a year, but by the time it broke six years later it seemed like a lifetime. Most of the stock had been sent out on agistment and much of it had been sold to provide an income. They were left with only 1,000 cows with which to rebuild the herd.

By the time the drought was broken by a sudden and torrential six inches of rain, followed by more rain a few weeks later, the country looked as if it was beyond repair. Indeed, some experts predicted that it would never become useful again whilst others, more optimistic, thought it might recover in ten years provided it was not stocked with cattle.

They were all wrong. Financially crippling though the drought had been, the country did not suffer as much as they

An experimental turbo windmill being tested at Undoolya.

had feared and when it was over the land showed a resilience that nobody had expected. Regrowth was good even in those parts which were restocked with cattle and slowly the herd grew, the land recovered, and Undoolya became a cattle station once more.

Indeed, Ted Hayes firmly believes that this land is better for being used: the virgin country that was barely able to provide a subsistence living has been turned into a highly productive station through years of careful management. The stock has improved too, for he has maintained the emphasis on quality started by his father when he ran Undoolya. They are continually upgrading the herd by ruthlessly culling the females and by importing bulls of only the highest quality.

The property is now run by Ted's sons: Bill, who lives with his family at Deep Well; and Jim, who runs Undoolya from the modern homestead which nestles beneath the red outcrops of the Jessie River. The original homestead that Bagot built is still in daily use as a kitchen.

Ted Hayes is now in his seventies but he is still rugged and active as befits a man who has spent a lifetime in Central Australia. He lives with Jean near the homestead at Undoolya in a comfortable modern house. The Albert Namatjira paintings on the walls, which immortalise the landscape of Central Australia, are simply faithful reflections of the country outside.

It is a country that Ted Hayes knows as few ever will and loves as few can. More than that, he has a sense of continuity with the past, together with a great admiration for those who battled against hardship and difficulties to make Undoolya what it is today.

'They must have had great hearts to have stayed and created the beginnings of what we now take for granted,' he says.

And no tourist would argue with that.

THE STATION COOK

It used to be said that the station cook had to be either a good cook or a good fighter. It is less true today perhaps, but the cook is still as important as he ever was for his skill (or the lack of it) will determine the well-being of nearly everybody on the station.

On a big property the cook might have to be very skilled indeed. Not only will he have to feed a large number of people every day, he might also have to send meals to the homestead (sometimes known as Government House) especially when there are guests.

'Pommie Bob' Foulston has been cooking at Brunette Downs in the Northern Territory for over twenty years. Born at Sheffield in England in 1924, he trained as a butcher before joining the Royal Navy for wartime service. He came to Australia in 1949 to escape the post-war austerity in Britain and after a brief time as a tobacco grower in southern Queensland he turned his attention to cooking.

He cooked for a number of shearing teams and at a hotel in Longreach before moving to the Northern Territory in 1955 as station cook at Helen Springs. In 1960, when at Oban Station, a friend of his who had recently taken over as manager of Brunette Downs asked him to join him there as cook. Bob Foulston arrived one Sunday, started work straight away, and did not have time to unload his car until three months later.

His day starts at 5.30 and his first job is to bake the day's supply of bread and rolls. He serves breakfast of steak, eggs, rissoles or sausages at 6.30 and spends the rest of the morning preparing later meals, making sausages or doing his own butchering. He uses about 500 kilograms of beef every week.

He serves smoko at 9.30 and a lunch of cold meat or salad at 12.30. Afternoon smoko is at 3.30 and at 7 o'clock he serves the evening meal of roast beef or hot corned beef.

He is a warm, cheerful person who over the years has been 'father' to countless jackeroos on this isolated station, many of whom write to him afterwards to thank him for his help. They do not even have to remember his address; 'Pommie Bob, Northern Territory' is all a letter needs.

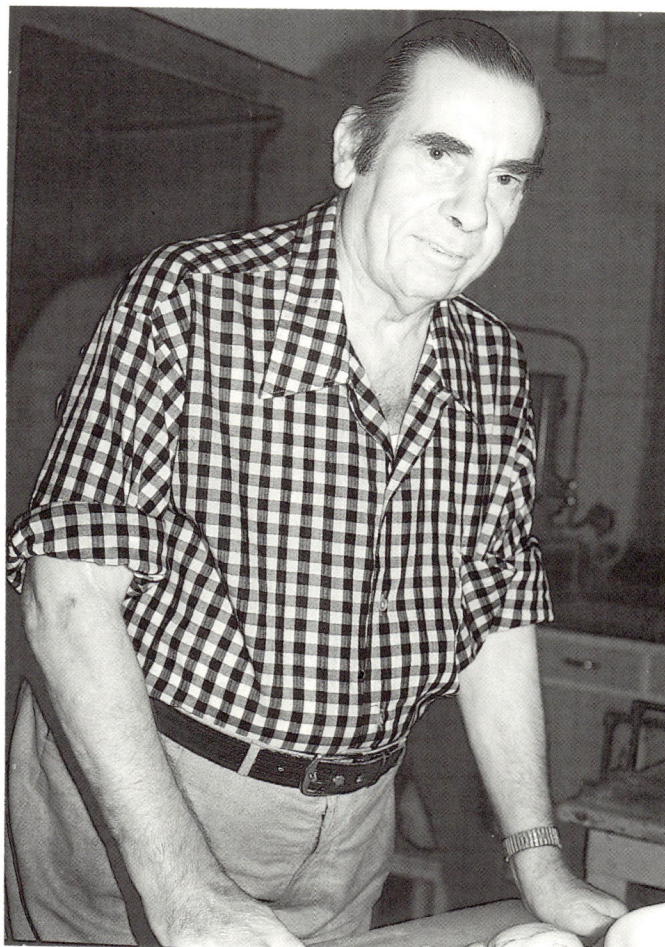

'Pommie' Bob Foulston, cook at Brunette Downs for more than twenty years.

Men working on stock-camps in distant parts of a property usually have their own cook with the camp. On a property working only one camp the station cook might go with them, leaving his wife to cook for the few people remaining at the homestead area.

Cooking on a stock-camp is vastly different to cooking in a kitchen. There is usually no refrigeration, supplies are limited to what can be carried in the trailer, and the cooking will be done on an open fire.

On the stock-camp at Innesvale in the Top End Terry Peterson gets up at 4 o'clock and lights his fire for breakfast. He then wraps up his supply of meat, which has been standing outside in the cool of the night, and stores it away for the day. If the camp is to be moved during the day he will start packing his gear on to the trailer after breakfast, but if not he will probably spend part of the day baking bread or making pastry, as well as preparing the evening meal.

The secret of baking on a stock-camp where the temperature can be over a hundred degrees is, according to Terry Peterson, quite simple: you keep all the ingredients at the same temperature. The water, for example, should be at the same temperature as the yeast, no matter how warm it might be. Here is his recipe for making bread in a camp oven.

Mix a tablespoon of dried yeast with 1.2 kilograms of flour, add half a tablespoon of salt and 1.2 litres of water. Punch the mixture into a round ball and then let it stand under cover for one hour, when it will have doubled in size. Punch it down with the heel of the hand to its original size, then leave it to rise again, which should take about thirty minutes. Punch it down once more, then put the mixture into a greased camp oven and put the lid on.

Meanwhile, dig a round hole in the ground so that it is 5 centimetres broader and 5 centimetres deeper than the camp oven. Fill the hole with wood, light a good fire and leave it to burn right down. Then remove the bottom layer of coals, put the camp over in the hole and cover it with these coals. Bake for one hour. If smoke starts to come through the coals on the top of the oven it means there is too much heat. Lift the camp oven out and let the hole cool off for ten minutes before returning it.

Terry Peterson with a loaf of bread baked on a stock camp in the Northern Territory.

NAPPA MERRIE

It is late on Monday morning and the winter sun is pleasantly warm as the road trains rumble out of the cattle yards at Mount Isa in western Queensland and head for the open road. There are three vehicles each pulling two double-deck trailers and a fourth pulling two single decks. On board are over six hundred Santa Gertrudis bullocks. A few days ago they left the property of Alroy Downs some 500 kilometres to the north-west and now, after a spell in the yards, they are starting out on the last leg of their journey.

Puffs of smoke burst from exhaust pipes above the cabs as the drivers move up through their multitude of gears. The huge vehicles, each nearly 50 metres long, slowly pick up speed and soon they are spaced out along the open road. Each driver settles down to a cruising speed suitable for his vehicle and before long the boss driver is 3 kilometres ahead of the single-decker at the rear. There is no sense of competition, no attempt to race, and little urgency, for they have 1,000 kilometres and many hours ahead of them. They are going to Nappa Merrie.

Far away in the south-west corner of Queensland, Nappa Merrie has a magic that is difficult to define but which is almost tangible. It is as if it demands a personal response from all who see it. As if, refusing to be accepted passively, it requires instead an active acceptance or rejection.

Perhaps it is its remoteness, for that is real enough. People there make light of it, of course, saying that there is another place 'further out' which is *really* remote. But this harsh and arid part of Australia is remote enough for most. Quilpie is 515 kilometres to the north-east, whilst their main service centre, Broken Hill, is 660 kilometres to the south and dirt road all the way. When creeks and rivers flood and cut the roads it may be impossible to reach either. Then there is no alternative but to wait with true bushman's patience for the water to go down. For, like the tide, it accepts no human intervention.

This rise and fall of the water, which is surely part of the magic of Nappa Merrie, is also the reason for its existence as a cattle station. With an average rainfall of only 7½ inches, and often none at all for years on end, this wilderness seems far too dry to support life in anything but its most basic or adaptable forms. It is desolate, huge and beautiful and in bad seasons almost hostile. Then, living things concentrate on survival, and not all succeed.

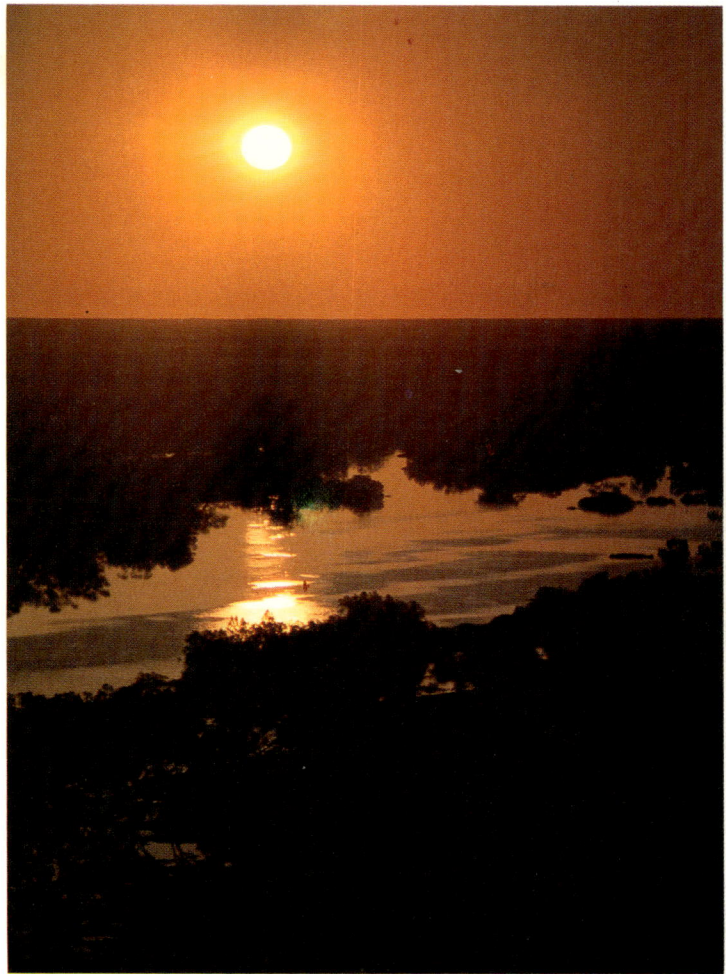

Sunrise at Cooper Creek.

But in the good seasons a miracle happens and it starts not here, on the banks of Cooper Creek, but hundreds of kilometres away in central Queensland. Heavy rain there will swell the rivers and make them run with a ferocity that is almost spiteful, often flooding the surrounding country so that travel becomes difficult and stock has to be moved to higher ground. The rain might last only a few days and although the benefits it brings can be vital, the memory of it recedes as the floods disappear and life returns to normal.

But the water, confined again to rivers and creeks, does not disappear so quickly. Instead, it makes its way south along the river system until it joins Cooper Creek, and that is where the miracle happens.

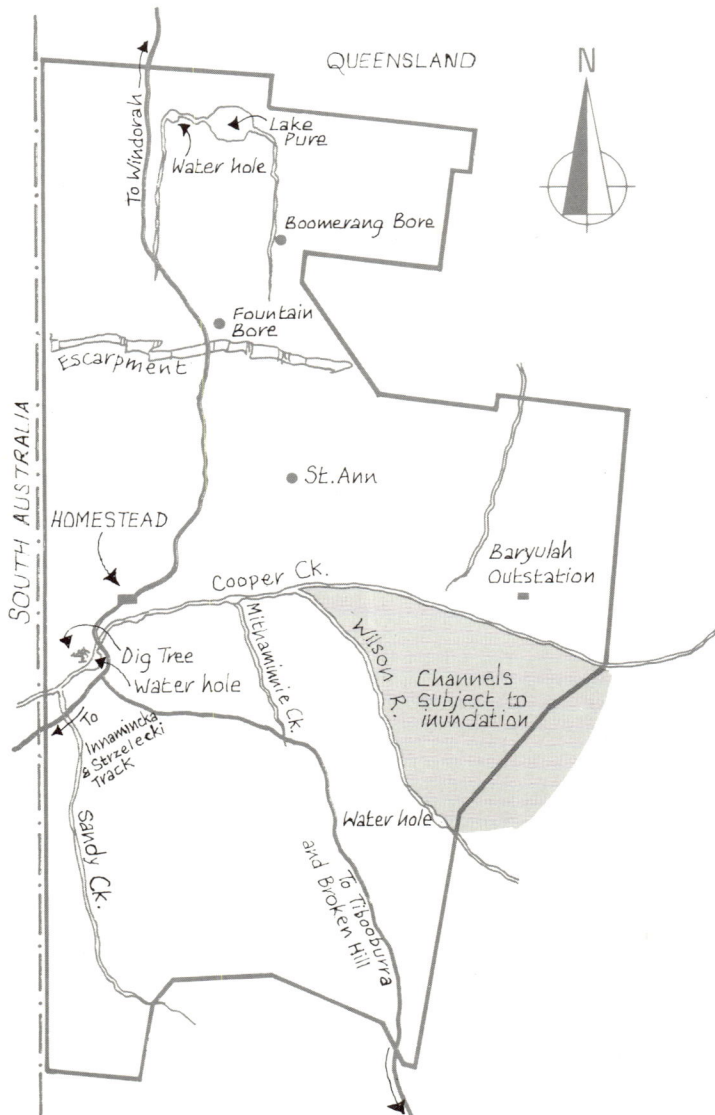

Drying wool at Nappa Merrie in the 1920s. The wool was laid on calico sheets and left to dry in the sun.

Nappa Merrie in the 1920s. The homestead in the distance was built by John Conrick in 1885.

Cooper Creek and its associated watercourses are often little more than dry sandy beds punctuated by waterholes which become smaller in the summer heat. The local rainfall is barely enough to maintain their level, let alone set the creeks running. But the heavy rain further north can. It might take weeks or even months for it to reach this Channel country, but when it does Cooper Creek starts to run again and as the water rises it finds its way into the countless channels that lace the low-lying country beyond.

Suddenly, in the middle of the desert there is water everywhere. The whole of the country between Cooper Creek and the Wilson River becomes a mass of rivulets which thread their way through the sandhills, the trees and the lignum like the veins of a huge leaf, before joining together to become Cooper Creek once more.

These rivulets make a natural irrigation system which puts water and fertile silt over ground that might have been dry for years. And when the water recedes it is replaced by a mass of new growth, vibrantly green and lush, that extends for hundreds of square kilometres and whose clearly defined edge against the red of the sand shows how high the water came.

It is an unpredictable miracle. The Cooper might remain dry for years before there is enough rain in central Queensland to fill its channels. But they are full now, and that is why the convoy of road trains has started the long journey to Nappa Merrie.

It is hardly surprising that European explorers failed to respond to the magic of this Channel country, for most were facing extreme hardship by the time they reached it. Most, indeed, were not concerned with reaching it so much as getting away from it.

The first to pass this way was Charles Sturt, who in 1844 left Adelaide on a journey to the centre of the continent which was expected to take two years. As he travelled north beyond what is now Broken Hill he became desperately short of water and in February 1845 he was forced to retreat when only a few kilometres from the then unknown Wilson River.

The following August, whilst still on the same journey, he discovered Strzelecki Creek near present-day Innamincka but it was not until he returned from the north two months later that he discovered the much larger watercourse nearby.

The Burke and Wills Dig Tree on the bank of Cooper Creek at Nappa Merrie, not far from the homestead.

An early photograph of station hands loading the Nappa Merrie wool clip.

He called it Cooper Creek after a South Australian judge. He made another attempt to travel from there to the centre of the continent but he was finally turned back by illness. He returned to Cooper Creek with the intention of following it to the east but after exploring only a few kilometres he prudently turned back for the long trip to Adelaide.

The next visitor was Augustus Gregory, who in 1858 led an expedition in search of the missing explorer, Ludwig Leichhardt. Lack of feed and water forced him to abandon his search at the junction of the Barcoo and Thompson Rivers in western Queensland. Reluctant to return, as he carried a good supply of stores, he decided to travel south to see if the Barcoo River eventually joined Sturt's Cooper Creek. He proved that it did when he arrived at Sturt's most easterly point. Continuing down Strzelecki Creek, he reached the Flinders Ranges and from there made an easy journey to Adelaide.

The next explorers were not so fortunate. In 1860 Burke and Wills left Melbourne at the start of the most famous, and perhaps least relevant, journey of Australian exploration. Their intention was to make the first crossing of the continent from south to north but the knowledge that John

McDouall Stuart was about to make a similar attempt from Adelaide turned it more into a race than a careful exploration of largely unknown country.

With their party already split by argument and bad temper, Burke and Wills reached Cooper Creek in November 1860. They established a depot on the bank of the creek and the following month Burke and Wills and two other men left to travel north to the Gulf of Carpentaria, leaving Brahe and the rest of the men to build a stockade at the depot and to wait their return.

The small party arrived within a few kilometres of the Gulf in February. Unable to go further because of swamps, they turned back to Cooper Creek. But what had been a race against Stuart on the way north now became a race to survive as they headed south. One man died on 17 April and four days later the other three, now barely alive, reached the depot at Cooper Creek.

Unfortunately Brahe, who had waited eighteen weeks for them, had left the depot earlier the same day to start his own return journey. He had sensibly buried provisions for them and had carved a message on a nearby tree telling them where to dig.

Two days later Burke, Wills and King left the depot, but instead of following Brahe back to Menindee they decided to follow Gregory's route down the Strzelecki Creek to the Flinders Ranges. They soon found they were too weak to travel and after a few weeks they were forced to return to the depot. Although they were well looked after by Aborigines, Burke and Wills died at the end of June. King, the only survivor, was eventually found in September by a relief expedition that had been sent out from Melbourne.

It was hardly the stuff to fire men with enthusiasm for Cooper Creek. But although Burke's competence as a bushman had been slight, especially when compared to Stuart (who successfully completed his journey and returned safely to Adelaide with all his men), the Victorians saw it differently. To them it was heroic (which is questionable) and tragic (which is beyond dispute) and they revelled in the drama of it.

One person was so stirred by these events that he decided he must see Cooper Creek for himself. His name was John Conrick and he was nine years old.

Born in Geelong in Victoria, he grew up on his father's property at Warrnambool surrounded by relative comfort. But the dream stayed with him and in 1872, when he was twenty, he was ready to take a mob of more than a thousand head of Shorthorn cattle and three friends on the long overland drive to Cooper Creek. Conrick had grown into a tall athletic man. The day before their departure he rode 16 kilometres to practice on a rifle range, then rode another 37 kilometres for a game of football and, having won, stopped at all the pubs on the way back.

The expedition was well planned, however, and Conrick's ability with cattle was beyond question. Covering about 14 kilometres a day, the party slowly followed the course of Burke and Wills until, 1,900 kilometres and eighteen weeks later, they at last reached Cooper Creek.

John Conrick and Robert Bostock now followed the Cooper west looking for the best land. Passing the still recognisable depot of Burke and Wills, they stopped at what they thought was the border of South Australia under the mistaken impression that Thomas Elder had taken up all the country beyond. But they were already too far west. Bostock decided to take up several blocks of land at Innamincka thinking that it was in Queensland, but when the border was surveyed six years later it was found to be in South Australia and he lost the land.

Meanwhile Conrick returned about 160 kilometres to where they had first joined the Cooper and decided to settle there near the Baryulah waterhole. He moved his share of the cattle on to the land and started to put up yards and buildings whilst Bostock travelled to Charleville to file their claims. Ironically, Bostock had no difficulty registering his own claim to Innamincka but he discovered that two other men had already claimed the land that Conrick had occupied, even though they had not even seen it.

Deprived of his first choice, Conrick decided to settle about 80 kilometres downstream at a place called Nappa Merrie. This time he went to Charleville himself to register it and then returned to Baryulah. A few weeks later the Aborigines attacked him in the early dawn and it was only the ferocity of his dogs that saved him.

When Conrick moved on to Nappa Merrie later in 1873 he was joined by George Ware, who had made his way from Melbourne behind the main party. Conrick was now low on supplies and as the level of the creeks meant that the route to the east was uncertain, he and Ware decided to establish a route down the Strzelecki Creek to Port Augusta. Although more than 900 kilometres away, the nearest port on the east coast was nearly twice that distance.

The route had been travelled before and when Conrick and Ware set out in 1874 they had some knowledge of it. In 1870 Harry Redford had come this way with a mob of a thousand head of cattle which he had stolen from Bowen Downs near Aramac in Queensland and which he had

John Rickertt, manager of Nappa Merrie, beside the station Cessna.

A road train convoy makes its way to Nappa Merrie.

Bullocks jump from the road train at the end of their journey from Alroy Downs.

The magic of the Queensland Channel country.

overlanded in an epic trip to South Australia. Conrick and Ware followed his route and succeeded in developing a stock route which remained in use for years afterwards.

On their return, Conrick and Ware continued to build up Nappa Merrie and their herd increased in a run of good seasons. By 1891 they were running over 5,000 head of cattle and about 500 horses on their original lease of 128,000 acres and Conrick had built a substantial homestead on the top of a big sandhill on the northern bank of the Cooper.

In 1891 they also acquired the neighbouring property of St Ann. Consisting of 1,209 square kilometres, it had been taken up in 1879 and had been used to run sheep. By the time it became part of Nappa Merrie it was carrying about 7,000 ewes and nearly 400 horses and a few head of cattle. A stone homestead had been built about the same time as that at Nappa Merrie to replace the original slab hut that had been built by the first owners.

Conrick and Ware were happy to run sheep as well as cattle on their bigger property, although they soon found that each had their problems. In 1899 the property was in a drought that had already lasted five years. When they sent their wool clip to Charleville on two horse-drawn wagons it took nearly two years to get there and they lost twenty-two horses on the way.

But by 1907, when Ware withdrew from the partnership, Nappa Merrie was well on the way to recovery. The drought was over and the property now carried nearly 10,000 sheep and 600 head of cattle. After a succession of owners, however, the land at Baryulah had not been so successful. The last owner, a member of the Victorian parliament, had forfeited the lease in 1904 and in 1908 John Conrick was finally able to add its 5,180 square kilometres to Nappa Merrie, thirty-one years after he had first applied for it.

In 1912 Nappa Merrie was running 6,000 head of cattle and had already developed the first commercial herd of poll Shorthorns, bred from a cow that had been part of the original mob overlanded from Victoria. That year they cut 405 bales from 16,000 sheep and the following year a huge wool scour was built not far from the homestead. Noisy and smelly, it was a steam-driven machine which removed the grease from the wool to make it lighter (and therefore cheaper to transport) and more acceptable to the distant buyers.

But it was to be the last good season for decades. The property depended on the floods that came down Cooper Creek but now it seemed as if this miracle might never happen again. Between 1913 and 1915 they lost sheep and cattle by the thousand in a drought that lingered into the Twenties and beyond. When John Conrick died in 1926, one of the forgotten giants of the early pioneers, Nappa Merrie was still in drought.

In 1929, under John's son Ted, the property lost 80 per cent of the sheep and nearly half the cattle as the drought continued. They had relinquished the property of St Ann in the drought of 1904 and now the only feed on the whole of Nappa Merrie was at Baryulah. It was a desperate struggle to

survive and there was still no relief. Although St Ann was reacquired in 1948 brumbies had destroyed the feed there and at Baryulah dingoes were so numerous that it was no longer possible to maintain the flock of sheep, small though it was by now.

The Conricks had owned and run Nappa Merrie for nearly eighty years. They had selected it, stocked it and lived on it. They had enjoyed success and endured hardship and they, above all, had known the magic of it. But now it was time to leave.

In 1955 Ted Conrick sold Nappa Merrie to its present owners, a family owned group which is now run by Trevor Schmidt, who is also managing director of the Australian Agricultural Company.

This family company also owns Bowen Downs in Queensland (from where Redford stole his huge mob) and Alroy Downs in the Northern Territory and because of this the role of Nappa Merrie could now change. Whereas the Conricks had to breed their cattle and sheep on the property, and were totally dependent on the seasons and the flooding Cooper for their success, the new owners did not need it for breeding.

Cattle could be bred more safely and reliably on their other properties and when the Cooper flooded they could be sent to Nappa Merrie to fatten quickly on the abundant feed that followed. If the Cooper did not flood, the cattle could remain on the breeding property. They would not grow so fat and they would probably be sold for a lower price, but they would at least survive and fetch something. And that was the difference.

Today, Nappa Merrie consists of 7,000 square kilometres of open country. Its western boundary is also the border between Queensland and South Australia and on the far side is the neighbouring property of Innamincka. Cooper Creek runs from east to west across the middle of the property and about half way along its length the Wilson River joins it from the south-east.

Further along the Cooper, not far from the western boundary, the Burke and Wills 'dig' tree stands as a reminder of their tragic and futile expedition. It is a lonely open plain dotted with coolabahs and with a line of lignum running along the water's edge. People have been married here, others have camped here for the Birdsville races, and others have pushed their four-wheel drives north from Broken Hill to spend a day or two beside this famous tree. But usually it is undisturbed, unchanging except for the seasons and the unpredictable water of the Creek.

There are three different types of country on Nappa Merrie. In the northern and southern areas, and to a lesser extent to the east, it is sand country. Typical of the desert of Central Australia, it consists of sandhills and smaller ridges that run from south-west to north-east.

Then there is gibber country: vast areas of undulating land completely covered by large, jagged stones. It was the

A stock camp at Nappa Merrie.

scourge of early explorers, for few animals could travel well on it, and it looks so barren as to be quite useless. But it grows a cover of Mitchell grass and with good local rain this cover can hide the gibbers completely.

Finally there is the river country, the key to the success of Nappa Merrie. It is associated entirely with the Wilson River and Cooper Creek and the myriads of channels that cover the country between them. It is well timbered and extensive but its usefulness depends entirely on the flooding of the watercourses. Whilst this country relies on natural water, the rest is watered by thirteen bores, ten tanks and five wells.

They run about 10,000 head of cattle at Nappa Merrie and most of them are Santa Gertrudis bullocks. They turn off about 4,500 each year, although this depends greatly on the flooding and the feed it produces. When bullocks are brought in they are usually put into the northern part of the property near Lake Pure. The lake is usually dry but patchy rain can fill the local creeks and provide a covering of feed for the incoming mob.

For the first few days they are yarded each night and allowed to graze under the control of stockmen during the day. After a few days, when the cattle have settled down after their journey, the stockmen ride off and leave them to look after themselves. After about six months they are mustered and driven south to the river country for their final fattening. When they are big and sleek and fat they are mustered again and put into the yards not far from the old homestead at Baryulah. There they are loaded on to road trains for the long journey to Quilpie, from where they are sent by rail to Brisbane.

Because Nappa Merrie is free of disease (as are the bullocks that are sent there) there is no need for a great deal of stock work. There are no calves being born, the bullocks are already branded and provided the feed is good they can be left to themselves for much of the time. Nor is there much need for the small paddocks which are necessary when disease has to be eradicated. The whole of the property is divided into only six paddocks. The largest, an almost incredible 2,500 square kilometres, extends south from Lake Pure, down the escarpment that runs across the property, and across the lower ground to the very edge of Cooper Creek.

The property is run by a staff of ten and even with families the total population is rarely more than fifteen. Of these ten, most make up the stock-camp which spends much of its time either at Lake Pure, Baryulah or on maintenance work. This is a small number for 10,000 head of cattle on 7,000 square kilometres and it is possible only because of the skilled and almost continual use of the station Cessna. Worth about four men, it is flown by the manager, John Rickertt. If his skill is undeniable — 'He can fly that Cessna through the eye of a needle,' another pilot once said — it has also been known to terrorise many of his passengers.

Born in Brisbane in 1947, John Rickertt went to school at Mackay before leaving at the age of fourteen to work as a ringer on a number of small properties owned by relatives. In 1967 he joined Stanbroke Pastoral Company as a ringer at Augustus Downs in the Gulf country, but was sacked after a punch-up with the head stockman. He stayed with Stanbroke, however, and spent the next few years working on a number of their properties. After eight months on Myroodah in West Kimberley he joined Vesteys as head stockman at Morestone Downs on the Barkly Tableland and in 1974 became manager of Oban, south of Mount Isa. In 1977 he was appointed manager of Nappa Merrie and has enjoyed its magic ever since.

'All through my working career I've found people I liked and some I didn't,' he says. 'I wouldn't run the best property in the world if I didn't like the people who owned it. But there is more to it than that. Properties and companies have to be run by practical men with practical experience. There is no room for amateurs in this business.'

He learnt to fly in 1972 after leaving Stanbroke and it was a turning point in his career. Aviation was gaining popularity in the industry at that time and he thought that learning to fly would not only advance his career but give him an alternative should he ever need one. So he took a long holiday at Coolangatta and had a trial lesson. 'It felt like a horse with a good mouth,' he said afterwards. He flew whenever he could and three months later, and $1,100 lighter, he had his licence.

He now averages about 750 hours a year in the Cessna and cannot imagine running the property without it. He has a mustering permit which allows him to fly below 150 metres, and that is where he spends most of his time. Mustering with a plane needs a high degree of skill as well as an understanding of cattle and it is usually done just above ground level with the plane swinging round the trees or rustling the top of the lignum. Checking fences might seem less hazardous, but only if you are on the ground. 'I usually fly along the fence at about 15 metres,' he says, 'but if I see something that needs fixing I go lower for a better look.

The homestead at Nappa Merrie still stands on the sandhill beside Cooper Creek. The solid, stone house built by John Conrick has been extended on the two narrow sides so that what were verandahs in his time are now passages with more rooms beyond. The big and comfortable house looks down to the Creek on one side and across a lawn on the other. There is a neat garden with rose beds and date palms and more radio masts than a tracking station. The radio room in the homestead is the only contact with the outside world and today Helen Rickertt is using it to locate the road trains.

Tall, fair-haired and still in her twenties, Helen Rickertt is bringing up her family in an isolation that would appall most people and which at first unsettled her. Born in

A stockman on an Appaloosa waters newly arrived bullocks.

The old homestead at St Ann.

Huddersfield in the north of England, she came to Australia with her parents in 1960. She went to school in Brisbane with the hope of one day becoming a doctor but when the money ran out she left in 1971 to work for a health fund. After moving to Mount Isa, in 1974 she visited her mother who was then station cook at Oban. She met John there and they were married the following year. 'I adjusted to the bush very quickly because I wanted to,' she says. 'I never expected him to give it up.'

By the time they moved to Nappa Merrie she was six weeks pregnant. She left for Brisbane five weeks before the birth and when she returned with eight-day-old Charley the temperature was 42°C. The dangers were obvious, and they did not diminish as time went on. Dehydration, the river, and the inevitable snakes were continual threats and although she was more familiar with them by the time Billy and Jill were born, they could never be overlooked.

The problems would be much greater without the Royal Flying Doctor Service at Broken Hill. A doctor flies into Nappa Merrie every month to hold a clinic and there are two radio sessions with the base each day during which she can ask for advice. Accidents and serious illness are more of a problem, but fortunately they are also rare. In an emergency she can be in radio contact with a doctor in Broken Hill in five minutes, but she would have to wait two hours for him to arrive. During that time she would be given help over the radio, but it would be a long wait.

The radio also provides contact of another kind. Every morning the women on neighbouring properties use it to have a ten-minute chat. Known to the men as the galah session, it is a fixed radio schedule that can be overridden only by an emergency. But by suburban standards the neighbours are that in name only. Innamincka, the nearest, is 48 kilometres away, Karmona and Orientos are each 96 kilometres away in different directions, and Durham Downs is 160 kilometres away. She might see one of them once a month in the winter, but hardly at all in the summer.

The mail comes once a fortnight and John has to fly to Orientos to collect it. With it comes a supply of bread from

Surface stones, called gibbers, make up part of the country at Nappa Merrie. They were the scourge of early explorers.

The homestead at Nappa Merrie.

Broken Hill together with other top-up stores that can survive the two-day journey. Bulk stores come by road from Broken Hill every six months and when John takes the plane there for service he returns with a load of eggs, bacon and butter to re-stock the deep freeze and cold room.

She would not change it now. 'I could never live in a town now. The emphasis there seems to be on different things with everybody wrapped up in their own world. Besides, I couldn't stand not knowing the neighbours.'

Tuesday morning, and none of the neighbours have seen the dust of the road trains. Then one, even more distant than the others, calls her up to say that he saw them about two hours ago. John pulls out the map and works out where they should be now, then decides to fly north to meet them so that he can give them directions for the last part of the trip.

After an early lunch he climbs into the Cessna and takes off from the sandy airstrip not far from the homestead. Turning on to a northerly course, he levels out at 75 metres and starts to roll a cigarette.

Thirty minutes later he is flying north along the Queensland border. He is higher now, scanning the ground for the unmistakable plume of dust that would indicate the position of the convoy, but there is no trace of it. Puzzled, he changes radio frequency and puts out a call.

'Incoming transport from Nappa Merrie mobile. Do you read, over?'

Static, then a faint voice with heavy noise in the background.

'Nappa Merrie mobile, this is transport. We're having fun and games at a washout on the road not far from Arrabury. Problems with the front trailers but we're working on it. Over.'

'OK. Be with you in a few minutes. Out.'

He changes course and soon he can see the stationary convoy spread out in a semicircle round the dirt road. Nearby is an old airstrip and after a low-level inspection he decides it is safe to land.

Water from the last rain has run across the sandy road and taken a section of it away to leave a slab-sided drop about a metre deep and almost the full width of the road. The danger is that in edging round this gap the less stable trailer might overturn. It takes very careful driving to get it through. And now they have to do the same with the vehicles behind. Far from anywhere, they have to solve this problem for themselves.

When all the vehicles are safely through, John draws a map on the ground to show the road train boss how to get to Lake Pure. It will take them three hours yet, but for the drivers and the bullocks it is the last stage of the journey.

John Rickertt takes off from the small strip and flies ahead to make sure the road has no more unpleasant surprises, then flies to Lake Pure and lands beside the stock camp pitched on the waterhole not far from the yards. Michael Keane, the head stockman, brought the camp up from the homestead a few days ago to get ready for the arrival of the bullocks. Now

Keeping the fire going at the stock camp.

he and John walk round the yards to check them once more, then they boil the billy and have a mug of tea with the men as they listen for the sound of the vehicles.

Then, faintly, they hear the first of them. It is a muted roar, rising and falling as the driver changes gear to cross the creek and changes again to pull his huge load up the bank on the other side. The ringers mount their horses and walk them round as the roar becomes louder, straining and throbbing as the vehicle grinds up the last hill. Then it comes over the top and the noise eases as the huge road train sweeps down the rise and stops between the camp and the yards, air brakes hissing as the twenty-two wheels at last come to rest at Nappa Merrie.

The road boss, an athletic man in his twenties, jumps down from the cab as the rest of the vehicles come over the hill. The noise is deafening, layer upon layer of it, then it dies to nothing as each vehicle sighs to a halt, one behind the other, at the end of a journey that has taken twenty-eight hours.

The ringers move up to the vehicles and their horses snort quietly and shake their heads so that bridles jingle and glint in the sun. And as the bullocks jump from the trailers and trot towards the waterhole, modern technology gives way to the much older skills of the stockman and his horse.

228

THE NURSING SISTER

In 1981 the manager of Brunette Downs had a problem. He needed a nursing sister and a bookkeeper on the station, but there was only one empty house. The solution, unlikely though it might be, was to find a husband and wife team, so he advertised in the Brisbane *Courier-Mail* for a nursing sister married to a bookkeeper who would be prepared to work on this isolated cattle station in the Northern Territory.

A few days later Jean Dryden, a district nurse at Bowen in Queensland, was looking for the racing pages in the *Courier-Mail* when she saw the advertisement. 'Howard,' she said, 'it looks as if we are going to Brunette Downs.' For Howard, her husband, was a bookkeeper.

Born in Tasmania in 1935, Jean Dryden trained at the Austin Hospital and the Royal Women's Hospital in Melbourne before joining the Freemason's Hospital there as a district nurse. She soon became interested in community nursing and in 1968, having by then married Howard, they moved to Burketown on the Gulf of Carpentaria, where she became matron of the local hospital.

'We went because it was different — and I wanted to see the Gulf. But when we got there I discovered that you can't see much of it for the mud and mangroves.' But it *was* different. One day she found herself delivering twins in the middle of a cyclone which blew the roof away before the birth was over.

As the nursing sister at Brunette Downs she looks after the Europeans who work on the station and the Aboriginal community which lives in a settlement not far from the station area.

There are about forty-five Europeans and most of her work with them, and the Aborigines who work on the property, is a result of accidents at work. Surprisingly these are fairly rare, but when they do happen they can result in severe lacerations or fractures that usually need immediate attention and which might require the patient to be moved to the nearest hospital, some 352 kilometres away at Tennant Creek, for more specialised treatment.

Snake and insect bites are more frequent. One jackeroo on the stock-camp was bitten on the toe by a huge brown snake which, they assumed, had spent the night

Jean Dryden, nursing sister at Brunette Downs.

with him in his swag. He recovered and, more surprisingly, overcame the experience and returned to work on the camp. Bites from spiders are even more common. Jean Dryden was herself bitten by a red-back when she was cleaning out an air conditioner.

Most of her work, however, is spent looking after the Aborigines, and because of this her salary is subsidised by the Northern Territory government as part of its own health programme. There are over a hundred Aborigines on the settlement and those not working for the station draw social security payments and support themselves from the station store. Jean Dryden runs the store and in this way she is able to see and talk to those Aborigines who might be reluctant to visit her daily clinic.

'My main job with the Aborigines is health education,' she says. 'That, and the care of children.'

The two are often very closely linked. Trachoma, for example, is an irreversible disease caused by dirt in the eyes, which eventually leads to blindness. It can be prevented in children by keeping their face clean, but the relationship between cleanliness in a child and middle-aged blindness is not always obvious.

Although she tries to send Aboriginal women to Tennant Creek to have their babies, most prefer to have them on the settlement with her as midwife. Usually the births are trouble-free, but if problems do occur the isolation is suddenly a real and additional danger. She can use the radio to consult the Royal Flying Doctor base, but she knows it will be at least two hours before a doctor can reach Brunette Downs and during that time she simply has to do the best she can.

But Jean Dryden, big, jolly and very capable, seems ready to cope with anything.

AUSTRALIA'S BROAD ACRES

Throughout its relatively short history Australian farming has been affected by many factors, but two have always had more influence than the rest and more often: the climate and the market.

Not that there is anything unusual in that. Since time immemorial farmers everywhere have looked at the evening sky and wondered what the weather would be in the morning; or trudged to market with a year's produce to discover whether they were to be rich or poor for the coming year. But in Australia these factors have been even more important than elsewhere.

The climate, unknown at first, was soon seen to be more hostile than most. Whilst the weather might not vary much from day to day in some parts of the continent, it was the longer variations that were the problem. Dry spells could turn almost imperceptibly into long and harsh droughts: then animals died and crops failed as the meagre supply of water disappeared.

Markets were, if anything, even less predictable. With a small domestic market that was often over-supplied even in bad seasons, it was necessary to look for bigger markets overseas. Those markets were subject to fluctuations of their own, but for the Australian farmer there was an additional complication — distance. This meant that anybody sending produce to these overseas markets in the first hundred years of Australian agriculture did so with no idea of current market values and had to be resigned to the fact that it might be a year before they even knew what the produce had fetched. As communications improved some of this uncertainty disappeared, but it was a slow process.

Farming in Australia started as soon as the First Fleet arrived in 1788. That motley collection of a thousand people and eleven ships had taken eight months and one week to reach Botany Bay. They brought with them about two years' supply of food and stores and had supplemented this with a small cargo of livestock which they bought at Cape Town during the voyage.

The settlers faced disappointment almost from the first day. Having been told that the soil at Botany Bay was fertile, they were appalled to find it poor beyond belief. And even after they moved to the more promising site at Sydney Cove they fared little better. Grain rotted in the ground, cattle were lost by negligent convicts and sheep were struck by lightning. By May the total livestock consisted of twenty-nine sheep, seven horses, nineteen goats, seventy-four pigs, two bulls, five cows and an assortment of poultry.

They also discovered that the country could provide them with nothing of its own. The Aborigines, about whom they knew almost nothing, were food gatherers, not food growers. In marked contrast to British colonies elsewhere, there was no agriculture to copy and no native labour skilled at growing crops. The result was that there was no local grain, no vegetables that they recognised and no milk-producing animals.

There was game and fish to add variety to an otherwise dreary diet, but rarely in sufficient quantity for them to replace other stores.

There was an urgent need, then, to grow food. The uncertainty of shipping meant that the first settlers had no idea when the next supply fleet might arrive and until it did they would have to survive on their own.

The settlements at Norfolk Island and Parramatta proved more fertile and were soon producing wheat and Indian corn. But fresh meat remained a luxury until the small number of livestock had increased sufficently to allow for some to be killed.

By 1803 the livestock had increased substantially. The government owned about 1,300 sheep and 1,800 cattle and another 650 cattle and nearly 10,000 sheep were owned privately. Five years later the government started to buy meat from private owners. As this practice grew, it was rivalled by a separate trade between producers and private buyers which was more profitable than dealing with the government.

But a drought soon reduced the number of animals and the colony of New South Wales turned to the even newer colony of Van Diemen's Land for supplies. Founded in 1803, that colony had soon become self-sufficient in meat and the new 'export' market came at a time when it was most needed.

This emphasis on meat production continued until the early 1820s and little attention was paid to other uses for livestock. Wool was still regarded as a by-product of growing mutton and in any case most of it was coarse and hairy. The government bought some for its mill which made rough cloth for convicts' clothing but the demand was too small to be very significant. When the price of wool rose in England in 1811 some better types were exported, but a fall in price some seven years later made this short-lived.

During the 1820s, however, free settlers started to arrive in some numbers and the population grew rapidly. Whilst this increased the demand for meat it also had another effect that was to have increasing significance. As these new settlers took up land many looked for livestock to put on it. The result was a healthy market for breeding stock which lasted until 1825, when financial difficulties in England cut off the supply of capital to Australia. This, together with a drought which started in 1826 and lasted for three years, brought the boom to an end and left many settlers unable to continue.

So far the settlement of land in New South Wales had been quite patchy. It had at first been concentrated around Sydney, Parramatta and the Hawkesbury and Hunter valleys, with the Blue Mountains forming what seemed like an impenetrable barrier to the west. This concentration of an increasing number of settlers in a relatively small area had inevitable results. The land was overgrazed and soon became tired, insects ravaged it, animal diseases increased, and the pressure to move out became irresistible.

When a route was found across the Blue Mountains in 1813 this pressure was relieved, although it was several years before settlement spread much beyond Bathurst. But by the end of the 1820s the movement outwards from the Cumberland Plain had gathered such momentum that many settlers had taken up land far beyond the reaches of government and particularly the surveyors, who were trying to keep some sort of order. The British government therefore defined an area called the Nineteen Counties which would be fully administered and in which it hoped to contain the development of new land until the outer reaches could be properly surveyed. Beyond the Nineteen Counties there would be no protection and no law.

Much of this development was based on cattle rather than sheep. In spite of the difficulties of importing cattle, especially in the early years, they had nevertheless become the most readily available livestock. The government had encouraged the spread of cattle and occasionally had even made grants of them to new settlers.

They had rarely been so generous with sheep. The demand for mutton could be supplied by small flocks that were fairly easy to keep, whilst the flocks needed to supply wool had to be much bigger. They were therefore restricted to those who had enough land and interest to run them and who, after the reduction of the import tariff in England in 1824, could handle the complexities of exporting to that market.

The depression that followed the drought in the late 1820s coincided with a fall in the price of coarse wool in London. But the price of better quality wool remained firm and pointed the way for those who could see it. This, and a reduction in the price of meat in Australia, led an increasing number of landowners to change to the production of fine wool instead of meat. By the middle of the 1830s sheep flocks had grown both in size and in quality, the merino was already established as a superb wool-growing animal and cross-breeding, albeit fairly unsophisticated, was well under way.

This, together with a new flow of capital from England, meant that more land was needed to accommodate the large flocks that now developed. Slowly men ventured further from the protection of the Nineteen Counties in search of land with good feed and reliable rainfall. Slowly at first, they sought and settled this new land and brought their sheep to places where no white man had been before. Making crude boundary marks, they simply took possession of it and settled down to an uneasy existence. They were called squatters, for that is what they did.

Much the same was happening in Van Diemen's Land. There, the good land between Hobart and Launceston had soon been occupied and with little room for expansion, men began thinking about the land near Port Phillip Bay. The Bass Strait, although dangerous, was by now well charted and had been used by sealers for many years.

Hume and Hovell had succeeded in reaching Port Phillip Bay in 1825 after an overland journey from Lake George near Goulburn in New South Wales but applications for land there had been rejected by the government because it was too far from other mainland settlements. But now the pressure in Van Diemen's Land was too great. In 1834 the Hentys crossed the Strait and took up land at Portland Bay and the following year John Batman led a group of settlers to Port Phillip Bay. By the time others flocked there from New South Wales the settlers from Van Diemen's Land had secured much of the best land and were already shipping their flocks across the Strait.

As this kind of movement could no longer be controlled by the government, Governor Bourke saw the need to give it some semblance of respectability and legality. The Squatting Act of 1836, therefore, gave any man the right to take up unoccupied land in return for an annual licence fee of £10 plus ½d for every animal he pastured on it. There was no limit to the amount of land he could occupy so long as he paid the fees and grazed sufficient stock on it.

Suddenly the pressure was released and in less than seven years nearly the whole of the fertile land from Brisbane to Melbourne had been opened for settlement. This massive surge outwards also created a spectacular demand for breeding stock and many found it more lucrative to meet that demand than to sell wool or meat. But as large areas became occupied, so this demand started to decline. At the same time the price of wool fell in England and this, coupled with a sudden loss of confidence there, led to a depression in Australia that started in 1841 and lasted for two years.

The result was disastrous. There was more meat than the colony could use, cattle became worthless, sheep could not be sold for mutton and the wool on their backs was hardly worth the cost of removing it. Fortunes in land were lost because runs only had value in terms of the

sheep or cattle they supported and when sold, the run itself was often 'thrown in' for next to nothing.

As the squatters tried to salvage what they could from the economic ruins, Henry O'Brien came up with the solution. At Yass in 1843 he discovered that sheep could be melted down for tallow. This, together with the sale of the skin and a little mutton, produced about eight shillings for each sheep. The following year the landowners in Victoria alone melted down nearly 10,000 sheep. As the practice became more widespread it decimated stocks, but the squatters survived.

When the economy started to improve in the mid-1840s interest in settling new land was revived, but now it was carried out with far more caution. With the closer land already occupied, this movement now spread into the Riverina, the Murray River, and the Darling Downs in Queensland. Other parts of Australia, meanwhile, had been settled slightly differently.

In 1829 a settlement had been established on the Swan River in Western Australia under the sponsorship of an English company. But the soil there was disappointing and it was some time before the settlement could be sure of surviving. And in 1836 another group of settlers, inspired by Edward Gibbon Wakefield, had left England to settle what is now Adelaide.

Wakefield's idea was that land should be settled in limited areas by those able to buy it. The money they paid would be used to bring out more people from England who would work as labourers until they had saved enough money to buy land of their own. Primarily based on cropping, the scheme was not a success and soon came to an end. By 1844 those running sheep and cattle in South Australia had started to move out of the selected areas to face trials of their own in the unpromising country further north.

In the east, though, new problems appeared as a consequence of the way the land had been settled. Those who had settled close to the centres of population had occupied their land on leases granted by the government. The government had stopped doing this in 1831 and had started selling freeholds instead. Now, as the towns grew and pastoral land on their edges became nore valuable, the earlier leases were cancelled and the land was sold for urban use. The leaseholder had the choice of either buying the freehold of the land at public auction or moving further out and squatting. The squatters, too, had no rights of ownership. So long as they paid the annual fees they had the right to use the land, but they did not own it.

Whilst all this had seemed reasonable at the time, the result now was that those using the land saw little point in trying to improve it. With no security of tenure, pasture improvement was unthinkable and few squatters thought it worth improving even the house they lived in. Many were still living in the crude huts they had built when they first arrived and women gave birth to children and brought them up in surroundings that were often primitive in the extreme.

It was clear that the traditional practice of putting into the land more than you took out would not be applied unless there was some security of tenure. The original system had been expedient, but a change was now needed. It came, after much public debate, in the form of an Order-in-Council in 1847 by which squatters were granted a fourteen-year lease on their land. After that they would continue to have preferential rights on any land they had improved. This provided the incentive needed to invest capital in the land, but it did more than that. It made the landholding into a property which had a saleable value.

A few years later gold was discovered. This had an even greater effect on the developing pastoral industry of Australia.

At first the effects were far from beneficial. Shepherds left their flocks, stockmen deserted their herds and shearers abandoned their blades as they joined the excited rush to the goldfields. But soon the benefits became apparent as people flocked into the colonies and the population soared. In the ten years from 1851 the population grew from 405,000 to well over a million, and they all had to eat. Once again the emphasis was on the production of meat and although the price of wool rose throughout the 1850s few bothered to take advantage of it. The profits from mutton and beef were much closer to hand.

At first the demand for meat was supplied from stock close to the goldfields but as this was depleted the benefits were felt over much greater distances. Great mobs of cattle were soon on the tracks from New England and the Darling Downs, heading towards the good land of the Riverina to be fattened for profitable sale in the Victorian goldfields.

Fortunes were made by those lucky enough to have good land in the fattening areas of the Riverina, but the boom did not last long. By the end of the Fifties the squatters in Victoria had increased the size of their herds and were able to supply most of the market. Overlanding was no longer necessary, and indeed became less profitable as the market stabilised and prices fell in the early Sixties.

The dramatic increase in population did much more than provide a demand for meat, however. Although some people left the colonies after the rushes ended, most stayed. Of these, many looked to the land as a means of investing their wealth if they had any, or giving them a living if they hadn't. So the demand for meat was followed by a demand for land, and this was less easily satisfied.

With much of the land, and certainly nearly all the best of it, now firmly in the hands of the squatters, a massive social movement gathered force. With a growing feeling of outrage, its aim was to 'unlock the land'. In a country that already prided itself on its equality, this force became unstoppable.

Although there was already enough evidence to show that farming in Australia was unlikely to succeed when carried out on a small scale, in 1860 and 1861 all the colonies passed Acts that allowed anyone to 'select' up to 640 acres of land and to establish themselves on it as cropping farmers.

As people rushed to find the best land, the squatters reacted with predictable fury as they tried to preserve their runs. The first step was fairly easy. As they also had the right to make 'selections', they promptly selected the best part of their own land. This always included the water frontages, which were then denied to those who selected the remaining blocks. Often this was enough to discourage other selectors and the squatter retained his land with little difficulty.

If not, he would make further selections in the names of members of his family, and when they ran out he used 'dummies' to act as selectors for him. If necessary a temporary hut was built, often on a corner where four blocks met, and a token cultivation carried out on a few acres. Even this was not always necessary as squatters were by now men of influence and government inspectors were rare and often amenable to 'consideration'. Some selectors found it more profitable to sell their selection back to the squatter as soon as they had acquired it, whilst others did so after a few years of backbreaking and unproductive work.

In the end the squatters who had been able to build up significant capital were able to withstand the siege. Many not only retained their runs but even increased them. Others less fortunate were unable to survive on their diminished runs.

With clearly established runs and firm ownership, the squatters were now able to develop their property with the knowledge that they would benefit from the investment. Fencing was often the first major improvement. Until now they had been makeshift affairs of logs, stone or anything else that came to hand, and the need for them had not been very great. But with an increase in animal diseases such as scab and pleuropneumonia the squatter needed to protect his herds and flocks from those travelling on roads along his boundary. The outer fence also defined his land, whilst those inside the property made it easier to manage the stock.

Fencing wire had been available since the 1850s but it had always been expensive. That was changed with the invention of the Bessemer process and soon steel wire became the standard material for fences.

This was improved even further in the 1870s with the introduction of barbed wire from America.

The invention in Victoria of the drafting gate also made it possible to sort sheep for the first time in a rapid and economical manner, in contrast to the English method of lifting them over the fence.

With wool prices buoyant throughout most of the Sixties, Seventies and Eighties, and with few periods of drought, many squatters found themselves affluent to a degree they had rarely imagined. It seemed that it would last for ever and it was time to revel in it. So they turned their attention to their living quarters, encouraged by families that could see little point in living in ramshackle single-storey homesteads now that they were rich.

Soon city architects were in great demand to design homesteads that would reflect the wealth and social standing of their owners. As each squatter tried to outdo his neighbour, huge Italianate mansions, often built on a scale that bore no relationship to his needs, started to dot the Australian bush. Typical was Werribee Park, which was built by Thomas Chirnside. Finished in 1878, it took five years to build and consisted of sixty rooms in two wings.

With his land now divided into paddocks, the squatter had to supply water to them. This was done by digging a well or building a dam. But he was still at the mercy of the climate until the discovery of the Great Artesian Basin under much of Queensland and New South Wales. This subterranean water solved many problems because although it was expensive to reach, the supply was much more predictable.

The first bore was sunk in New South Wales in 1880 and a few years later they were widespread. Soon they were being sunk wherever the geological structure indicated a possible artesian basin and they altered the whole possibility of running stock in the drier parts of the eastern States. Once again great herds of cattle and flocks of sheep took to the roads as stock was moved into these new areas which had been made available by this hidden water.

A major achievement in a different form added to the feeling of optimism. For many years scientists had been experimenting with refrigeration and the problem of maintaining insulated rooms at low temperatures. The results were not encouraging until the 1870s when Professor Linde of Munich made a successful machine that used compression. Almost immediately further experiments were made in Australia to apply this process to the holds of ships. Although the difficulties were immense, the benefits would be dramatic. Not only would Australia be able to ship frozen meat to Britain, but meat could be kept in cold storage in times of glut for sale later when prices improved.

Although it was a long time before all the problems were solved, the development of refrigeration had a profound effect on the cattle industry, especially in Queensland. The cattle population there rose from about three million in 1880 to nearly seven million sixteen years later and Queensland emerged as Australia's leading producer of beef.

But if the squatters thought that their problems were over, they were wrong. And if they thought that they alone were responsible for the success of pastoral Australia, they soon found they were wrong about that as well. The industry now involved a great number of people beside the landowners. Drovers moved cattle, bullockies brought in supplies and took away the wool clip, and shearers moved from one station to the next to bend laboriously over thousands of sheep for weeks on end. They were all part of the pastoral industry and the squatters needed them as much as they needed the squatters.

By the start of the 1890s, however, many squatters were in serious financial trouble. As prices fell, many found it increasingly difficult to meet their high mortgage payments, mortgages that had been run up in better times at high interest rates to finance the wave of improvements. The mortgage took priority over all other payments because if it were not met there was a real risk that the land would be lost. If times were bad for the squatter, as they now were, at least they could cut back on wages. Or so they thought.

They announced a reduction in the rate paid to shearers in the belief that there would always be plenty of men willing to work for whatever they could get. Shearers had organised themselves more or less informally since the 1880s but the movement had gathered pace until in 1889 the Amalgamated Shearer's Union had 20,000 members and controlled nearly 2,500 sheds. Faced with this reduction in rates, their growing militancy came to a head in Queensland in 1890 when the Union announced that its members would not work alongside non-union labour.

When the owners of Jondaryn Station on the Darling Downs started to shear with non-union labour, the waterside workers in Queensland declared the wool clip black. The station gave in, but it was only the beginning, not the end. A few weeks later the Union tried to extend a similar rule to all States and this was fiercely resisted by the squatters.

The result was a shearers' strike that was marked by threats and violence on both sides and which generated much bitterness. Threatened violence led to the arrest of some of the strike leaders in Queensland in April 1891 and most were sentenced by a hostile judge to three years hard labour. Four months later, with many of its members close to starvation, the Union finally gave in and agreed that its members would work alongside non-union labour. But it had not been in vain. From this strike developed a system of arbitration which in turn led to a steady improvement in wages and conditions for those who worked on the land.

For the triumphant squatters it was a short-lived victory for they were now to learn, and not for the first time, that after the good times come the bad.

The optimism that had generally prevailed throughout the Eighties started to change and two years later the Australian economy went into a severe depression that lasted several years. After years of unrestricted speculation, the whole financial system suddenly collapsed. Banks and building societies failed, graziers could not pay their debts and mortgages were called in. With their dreams of becoming the aristocracy of Australia shattered, many squatters were ruined and in the process many of the big runs, especially in arid areas, fell into the hands of companies.

When the depression came to an end in 1894 the squatters who had survived were changed men and many had learned a great deal about the impermanence of wealth that comes from the land. It was followed the next year by the start of a widespread drought that was to last, in increasing severity, until 1902. Waterholes dried up, feed disappeared and stock died by the thousand. The sheep population in New South Wales fell by half and the survivors competed with the rabbits for the last blade of grass.

Rabbits had been a problem for years. Introduced in 1859 at Barwon Park near Winchelsea in Victoria, they soon bred in millions and overran neighbouring properties. By the end of the 1860s they had spread through the Western District, the Wimmera and the Mallee and by 1872 they had reached the Murray. Fourteen years later they had reached Queensland and by 1894 they were at the border of Western Australia.

The effect was serious in some areas and disastrous in others, especially when drought made good land arid. Even in good times rabbits could reduce the carrying capacity of the land by half. In bad times they reduced it almost to nothing. The problem remained unsolved until the introduction of myxomatosis by CSIRO.

So Federation in Australia dawned on a more than tattered pastoral industry that was financially insecure and ravaged by drought. True, the price of wool was increasing and so were cattle herds in some areas in anticipation of attractive export business. But it was also clear that the industry had overreached itself by occupying marginal land and a slow retreat was already taking place. As the industry re-formed in the early years of the new century it was helped by two new factors, each of which was to have a real influence, and indeed still does.

The first of these was the use of superphosphate. It had been used by Victorian dairy farmers as far back as the 1870s, and in the 1880s and 1890s its value had been demonstrated on a bigger scale by the wheat growers of South Australia. The benefits were noticed by graziers, who now started to use it as a means of improving their pastures. The results were often dramatic, although it took many years to work out the best methods for different circumstances.

The second was the use of subterranean clover. This plant had been introduced into Australia accidentally and although it was not widespread some farmers had already realised its value. A nurseryman called Howard, who had a property in the Adelaide Hills, was the first to draw attention to it. From 1910 onwards he did a great deal to publicise it. Its value was that it could be grown as a pasture plant in winter rainfall areas that received less than 20 inches of annual rain. Whilst of great value in mixed farming, it also enabled pasture to be grown for sheep and cattle in the drier areas. In well-grazed areas it could revitalise pastures when used with superphosphate.

After a period of low prices, which in turn curtailed any rapid development, the First World War brought dramatic changes. Faced with a very high demand in traditional export markets, the Commonwealth government introduced marketing authorities to handle wool and dairy sales overseas. As orders came in, prices rose sharply and money once again flowed into the industry. This was the first it had seen for a considerable time.

There were other consequences that were even more farreaching. As young men left the land to fight in Europe and elsewhere, women had to take over their jobs. This led to a more willing acceptance of those machines that were already in existence, and created a demand for the development of new ones. The war also gave many young men an opportunity to see pastoral methods in use overseas and when they returned they brought with them a willingness to change established techniques which, with the growing awareness of the value of science, did much to revitalise the Australian industry.

It was soon needed. During the 1920s Australian pastoralists could look back on more than a hundred years of achievement, and during that time a great deal of experience had been acquired. They now knew much more about their environment and were learning to adjust to it instead of trying to impose inappropriate techniques on it. But in the process they had introduced some elements (such as the rabbit) which in turn threatened the whole ecological balance. Science had made them aware of it, and they looked to science for the solutions.

An example was the prickly pear. It had been introduced in the nineteenth century and by 1900 it covered more than ten million acres. It spread even more quickly during the drought. By 1920 it covered sixty million acres and was spreading at the alarming rate of a million acres each year. It could be controlled by the use of chemicals but much of the land it had infested in Queensland and New South Wales was worth less than £1 per acre and could not justify the use of costly techniques. The answer was found in 1926 with the introduction of a moth from Argentina. The caterpillar of the moth attacked the interior of the plant and as numbers increased they rapidly liberated almost the whole of the infected land.

Prices fell in the Thirties as overseas markets were caught up in the world-wide Depression. The Depression had a deep effect on Australian financial insitutions as well. Once again the fear of foreclosure was felt. Meat prices remained fairly firm, however, and many properties diversified to take advantage of them. By the mid-Thirties, 30 per cent of Australian sheep were to be found on wheat farms, where they were kept for the production of fat lambs.

The outbreak of the Second World War once again did much to revive the economy, but it also threw overseas markets into complete disarray. The Commonwealth government immediately offered to supply meat and wool to Britain at fixed prices. It then established targets in Australia for their production and at the same time introduced restrictions on their domestic price. Eventually the entire wool clip was taken over by the government for sale to Britain and the prices, as for those of meat, were significantly higher than they had been before the war.

But the aftermath was far from pleasant. The last two years of the war were marked by a severe drought that extended across the whole of Australia and with manpower reduced there was little that could be done to alleviate it. Livestock had been built up to meet the demands of war but fodder reserves were almost non-existent. There were massive rabbit plagues, the wheat crop failed and, inevitably, sheep died by the million. When men returned from overseas they found properties that were almost derelict and the means of restoring them in short supply.

Wool auctions started again in 1946 with the job of selling not only the current clip but the stocks that had accumulated during the war. Surprisingly, this proved easier than expected and prices were good. The outbreak of the Korean war in 1950 gave the market another boost and the average price jumped from the 1945 contract price of 15d per pound to a 1950 high of 144d per pound.

As a result money flowed once more into sheep properties and much of it was used to make much needed improvements and to purchase adjoining properties. With supplies more readily available there was a concerted effort to rebuild fences, sink bores and build new yards and sheds. Pastures were resown and stock increased, all in an incredibly short time.

Beef producers also shared in this new prosperity, although for different reasons. An agreement was made with Britain in 1952 under which that country would take 90 per cent of Australia's surplus beef and make deficiency payments should the price fall to an unprofitable level. When this happened in the mid-Fifties it coincided with a serious shortage of beef in the United States and Britain readily agreed that Australian beef should be diverted to this new market.

Meanwhile sheep numbers continued to increase until they reached a peak of 180 million in 1970, after which they declined due to droughts and falling wool prices. The effects of sharply fluctuating prices were well known on Australian properties, where the pattern of boom followed by bust was only too familiar. But now competition from the new man-made fibres meant that there was no wool boom to get the industry out of trouble as it had in the past.

The Reserve Price Scheme for wool was introduced in an attempt to iron out some of this unevenness. Under it, the Australian Wool Corporation fixes a floor price for all grades of wool each year. If an offering fails to reach that price at auction, it is bought at the floor price by the Corporation, which then stores the wool for resale when prices improve. The grower is therefore assured of a minimum price for his clip and from this pays a levy to the Corporation to finance its activities. He also shares in any profits that are made when the Corporation sells its accumulated stock.

Although less predictable, beef growers have benefited from the opening of new markets, particularly in Japan and other Asian countries. Lucrative though these markets are, they are often subject to political influences beyond the control of beef producers. Many importing countries fix quotas for each supplying country and these often have more to do with balance of trade than the quality of the beef.

Clearly the Australian pastoral industry has come a long way from its humble beginnings less than two hundred years ago and its arrogance of a century later. It is now the world's largest supplier of apparel wool and one of the largest producers of beef. In the process it has helped to support the entire Australian economy and indeed for a long time it *was* the economy. Even now, with mineral resources taking much of the glamour, Australia would not survive long without the pastoral industry and its export earnings.

But it has done much more than produce money. It has produced much of Australia's history and provided it with many of its legends, if not its identity. The people who made it happen — the squatters, the selectors, the drovers, the shearers and all the others — now seem slightly larger than life. But they remain part of us all.

BIBLIOGRAPHY

The following are some of the sources used in the preparation of this text which might prove useful for those wishing to enquire further.

About Wool, Series of pamphlets published by the Australian Wool Corporation, Melbourne.

Alexander, G. and Williams, O. B. (Eds.), *The Pastoral Industries of Australia*, Sydney University Press.

Austin, H. B. *The Merino: Past, Present and Probable*, Grahame Book Company, Sydney, 1950.

Barnard, A. (Ed.), *The Simple Fleece*, Melbourne University Press, Melbourne, 1962.

Bean, C. E. W., *On The Wool Track*, Angus & Robertson, 1910, reprinted Sydney, 1969.

Beattie, W. A., *Beef Cattle Breeding and Management*, Pastoral Review, Sydney, revised edition, 1971

Bolton, G. C. and Pedersen, H., *The Emanuels of Noonkanbah and Gogo*, Journal Royal Western Australian Historical Society, Vol. 8, Part 4, 1980.

Bonsma, Jan C., *Santa Gertrudis Past Present and Future*, Santa Gertrudis Breeders International, Texas, 1965.

Cameron, C. W. M., *F. F. B. Wittenoom*, Primary Industry Committee of the Western Australia 150th Anniversary Board, Perth, 1979.

Cannon, Michael, *Life in the Country*, Nelson, Melbourne, 1973.

Casey, Lord, *Australian Father and Son*, Collins, London, 1966.

Cole, V. G., *Beef Production Guide*, New South Wales University Press, Sydney, 1982.

Dickens, Jonathan, *Diary October 1882–December 1898*, Manuscript held at Terrick Terrick.

Dolling, C. H. S.. *Breeding Merinos*, Rigby, Adelaide, 1970.

Duncan, R., *The Northern Territory Pastoral Industry 1863–1910*, Melbourne University Press, Melbourne, 1967.

Durack, Mary, *Kings in Grass Castles*, Constable, London, 1959.

Falkiner, Suzanne, *Haddon Rig — The First Hundred Years*, Valadon Publishing, Sydney, 1981.

Feeken, E. H. J. and G. E. E. and Spate, O. H. K., *The Discovery and Exploration of Australia*, Nelson, Melbourne, 1970.

Foundation Souvenir Book, The Queensland Merino Stud Sheep Breeders Association, Brisbane, 1935.

'Haddon Rig', Notes in archives at Warren Public Library.

Harmsworth, T. and Day, G., *Wool and Mohair*, Inkata Press, Melbourne, 1979.

Hawker, Walter, *Reminiscences of George Charles Hawker of Bungaree*, copy held in South Australian State Archives.

Hewat, Tim, *Golden Fleeces* [History of Boonoke], Bay Books, Sydney, 1980.

Hill, Ernestine, *The Territory*, Angus & Robertson, Sydney, 1951.

Kelly, J. H., *Beef in the Northern Territory*, A. N. U. Press, Canberra, 1971.

Lilley, G. W., *The Story of Lansdowne*, Lansdowne Pastoral Co, Ltd, Melbourne, 1973.

'MacAnsh Family', Research notes, Mitchell Library, 1973.

Makin, Jock, *The Big Run*, [History of Victoria River Downs], Rigby, Adelaide, 1970, reprinted 1983.

Munz, H., *The Australian Wool Industry*, Cheshire, Melbourne, 1964.

O'Loghlen, F., *Beef Cattle in Australia*, Johnston, Sydney, 1948.

Pockley, L. A., *A Handbook for Jackeroos and Others*, Bay Books, Sydney, 1982.

South Australian Merino Flocks, Publisher unknown, 1936.

Thomis, M. I., *Pastoral Country — A History of the Shire of Blackall*, Jacaranda Press, Brisbane, 1979.

Wadham, Samuel, *Australian Farming 1788–1965*, Cheshire, Melbourne, 1967.

Weelhouse, F., *Digging Stick to Rotary Hoe*, Cassell, Melbourne, 1966.

Willey, Keith, *The Drovers*, Macmillan, Melbourne, 1982.

Wittenoom, Frank, *Memoirs from Murchison Pastoral and Goldfields Areas*, Geraldton Historical Society, Geraldton, no date.

Wright, Phillip A., *Memories of a Bushwhacker*, University of New England, Armidale, no date.

Ziegler, Oswald, *The Australian Merino*, New South Wales Sheepbreeders' Association, Sydney, no date.

INDEX